Digital Hardware Testing: Transistor-Level Fault Modeling and Testing

For a complete listing of the Artech House Telecommunciations Library,
turn to the back of this book

Digital Hardware Testing: Transistor-Level Fault Modeling and Testing

Rochit Rajsuman

Artech House
Boston • London

Library of Congress Cataloging-in-Publication Data

Rajsuman, Rochit.
    Digital hardware testing : transistor-level fault modeling and
testing / Rochit Rajsuman.
        p.   cm.
    Includes bibliographical references and index.
    ISBN 0-89006-580-2
        1. Electronic digital computers--Circuits--Testing--Data
processing.   2. Integrated circuits--Very large scale--Testing--Data
processing.   3. Fault-tolerant computing.   I. Title.
TK7888.4.R35   1992   92-8800
621.39'5'0287--dc20   CIP

British Library Cataloguing in Publication Data

Rajsuman, Rochit
    Digital Hardware Testing:
    Transistor-level Fault Modeling and
    Testing
    I. Title
    621.3815

    ISBN 0-89006-580-2

© 1992 ARTECH HOUSE, INC.
685 Canton Street
Norwood, MA 02062

International Standard Book Number: 0-89006-580-2
Library of Congress Catalog Card Number: 92-7350

10 9 8 7 6 5 4 3 2 1

This book is dedicated to scientists, engineers, and anyone else who seeks knowledge and keeps politics out of that search.

Contents

Preface

Generally, one hears that writing a book is an extremely difficult job. Popular belief is that many, many long hours and weekends are needed to finish a book in a timely fashion. Somehow, I was dubious and now I am glad that I did not believe it. I have seen many friends who did not initiate a book writing project because of this belief and sometimes because of a little discouragement from an "old pro." Also, by saying a job is extremely difficult, one may get a little extra credit and respect from the community. There is probably no harm in that (I would like to do so!), as long as first-time writers do not become discouraged. So, I have decided to tell of the actual effort in writing this book.

During my PhD studies, I found some myths regarding certain concepts related to VLSI testing. The idea of writing a book developed during a walk around Case Western Reserve University (CWRU) campus on a summer afternoon in 1990. Because of another research project, I could not do anything during the summer. In the fall, I prepared a proposal and started talking to the publishers.

From the start, I tried to measure the time and effort, but I cannot assign a number to the time spent during my PhD work, where I became acquainted with various topics related to VLSI testing. At that time, I read almost all the papers from the major conferences and journals (the list is given in Appendix B). So, when I prepared the proposal, I had a clear-cut idea about what to write and in how much depth. In the proposal, I also sorted out the sequence of topics and prepared three sample chapters for the publisher. These sample chapters were prepared in a very rough form, intentionally, so that I would receive a critical review. I estimate the total time for proposal and sample chapters as about two months, while performing all my other duties as normal. One may consider that as a full-time job for a month.

The review of the proposal and sample chapters took much more time than the writing itself. For about six months, I sat and waited. The reviews from one publisher were a little surprising and much more negative than I had anticipated.

The reviewer firmly believed that a comprehensive book covering most of the topics could not be written in six months. The publisher was still willing to take a chance, but, due to some other logistical problems, that deal did not go through. When I was looking for a publisher, I accidentally came across the Artech House, Inc. The Artech House staff treated me in such a wonderful manner that I immediately dropped correspondence with everyone else. This whole process took about six months. During this period, there was no work except for some phone calls or filling out a few requisition forms.

The actual writing did not start until June 1991. Before the summer ended, I had reworked the original sample chapters and written two more chapters. In August, when classes started, I scheduled my time to avoid any conflict. I made a policy to work only during regular working hours. I also kept my door open to the students and allowed them to stop by at any time to discuss anything from VLSI to UFOs. This open door policy spoiled the students, but at the time it gave me a break from the work. With all this and my normal duties (although, I could not publish the usual number of papers during this period), I had no problem whatsoever in finishing the whole manuscript by mid-December.

Based on this experience, my advice to anybody considering writing a book is that it does not matter if you are not very serious at this time; prepare an outline of the contents and contact a publisher. The scientific community needs good books. Being in academics, I know; I need good books. Pay no attention if someone says that it is a difficult job and you might exhaust yourself.

Having related this story, let me explain the organization and contents of the book. The first four chapters are devoted to the fault models and complexity of the testing problem. The reason for devoting four chapters to this topic is that it is poorly described in existing books. The complexity and testability analysis are needed in the beginning, but it has been given no consideration in the existing literature. Similarly, the bridging faults and open faults are the most important failure modes, but none of the books covers them. Thus, I was essentially forced to write separate chapters on these topics.

The second part of this book, Chapters 5 and 6, are devoted to the testing of combinational circuits. Chapter 5 includes a general test method applicable to random logic and Chapter 6 includes the testing of PLAs. Although a PLA also may implement a finite state machine, Chapter 6 is restricted to combinational testing. In these chapters, I also tried to include topics that are not covered in other books (i.e., algorithm FAN, switch level test generation, and the testing of EEPLAs).

Chapters 7 to 10 can be considered as the third part of this book, which is devoted to the testing of all types of sequential circuits, with or without extra hardware. Whereas memory testing is one of the most important topics, it is somehow left out of most books. To compensate for this deficiency, Chapter 7 is devoted to memory testing. Similar reasons apply to Chapters 8 and 9. All existing books include no more than one subsection (describing the checking experiment) on the testing of sequential circuits, while microprocessor testing is virtually ignored. To fill this gap, Chapter 8 is devoted to the testing of sequential circuits and Chapter 9 to the microprocessor testing. Chapter 10 includes the design for testability and built-in self-test methods. This chapter describes all the major testing techniques that use extra hardware. It may appear to be a concise chapter, but it indeed provides complete information.

Chapter 11 covers a new topic, IDDQ or current testing. Apart from brief mention, this topic has not even been referred to in any other book. Chapter 12 covers special test methods related to reliability. These test methods have been used in the industry for a long time. No shipment is made without completing reliability testing, but no book even mentions this topic.

I have tried to capture the basic knowledge in the area of IC testing and include all the major techniques. One topic that I could not include is testing for timing or delay faults. The delay fault model is included in Chapter 2, and Appendix B lists a few basic papers on this topic. One may also find that in some places I have not mentioned a program or the implementation of a particular algorithm. Sometimes, this was intentional, but otherwise due to limited space. However, I tried not to neglect the basic algorithm. I hope readers will be satisfied and able to find most of the concepts in these pages.

ACKNOWLEDGEMENTS

There are many people to whom I should express my thanks. First of all, I wish to acknowledge the Artech House staff members for the wonderful treatment that I have received. Special thanks are due to Pamela Ahl, Mark Walsh, and a former staff member, Rebecca Warren. I am also thankful to the Artech's reviewers as well as to other publishers whom I contacted. Reviewers' comments were very useful and they also identified some typographical errors. I also want to acknowledge Dhiraj Pradhan, Parag Lala, Warren Debanny, Sreejit Chakravarty, Yashwant Malaiya, Daniel Graham, Kedong Chao, and Bidyut Gupta for reviewing parts of the manuscript at my request. Their comments and suggestions were invaluable. I am also thankful to my co-authors Kamal Rajkanan and Sandeep Gupta for providing excellent contributions in a timely fashion. My thanks also go to Tushar Gheewala and Susheel Chandra for providing some material included in this book. I am also indebted to the IEEE Press for permission to use numerous diagrams. In the beginning of the project, some colleagues were a little reserved. I am really thankful to them because that greatly motivated me. Last, but not least, I am thankful to some of my graduate students, who, from time to time, pointed out some typographical errors.

Chapter 1
Introduction to Digital IC Testing

1.1 INTRODUCTION

Since the invention of integrated circuits (ICs), human life has changed dramatically. Every aspect of present-day life has been influenced by and become dependent on digital electronics. This influence of electronics on daily life became possible only because of the low cost, high accuracy, and reliability of electronic components as compared to their mechanical counterparts. The use of ICs in medical instruments and military and space applications became possible only due to accuracy and extremely high reliability. As the use of electronic components increases, the expectation of lower cost, better accuracy, and higher reliability increases. The IC design process tries to fulfill these expectations by putting more transistors per unit area of silicon, using design automation to reduce the design time, increasing the circuit operating speed, and reducing its power consumption. These design steps provide lower cost and better accuracy, but cannot guarantee reliability. In fact, as the circuit density increases, the probability of a manufacturing defect increases. The higher expectation of reliability can only be met by more thorough and comprehensive testing of ICs.

IC testing can be performed at different levels of abstraction. The complexity of fault detection experiments varies with varying degrees of circuit abstraction. The objective is to find manufacturing defects, which cause a fault and hence failure, or a potential fault or failure. In general, the detection of all existing and potential faults is not feasible. The normal procedure is to assume a fault model for the given circuit description and to test whether such a fault exists. Obviously, higher fault coverage provides better reliability. However, higher fault coverage results in the rejection of more ICs, and hence a lower manufacturing yield. From an economic viewpoint, an engineering decision is required to decide how much testing effort is necessary. Some basic concepts related to testability measures, fault coverage, and the complexity of test generation are essential to make such a decision wisely.

1.2 TESTING PROBLEM AND CONSIDERATIONS

Any electrical circuit is prone to two basic physical defects, short circuits and open circuits. At the circuit board level, one may use a large number of probes to facilitate the test. At the chip level, physical probing is limited to the input-output (I/O) pins. In this situation, not all physical defects can be tested. An alternative is to test for the functionality of the chip. If a defect is not causing a functional error, one may consider it irrelevant. For a given circuit, one can calculate the expected output for the given inputs. When this input is applied to the circuit, a fault-free circuit should produce the same output as the calculated (expected) output. If any bit in the output differs, the circuit has a fault. Note that here we seek only fault detection and not fault diagnosis, such as where the fault is and what the cause is.

To test all possible functions, we need to apply all possible input combinations while observing the output response. Unfortunately, this results in a very long test sequence, which is not practical. This point becomes clear by considering a simple example, such as a 32-bit adder. A 32-bit adder has 64 inputs. To test it exhaustively, we need to apply 2^{64} binary combinations. Even if we apply these combinations (test vectors) at a rate of 10 MHz, we need about 58,500 years! Obviously, such testing is not practical.

The problem worsens when we realize that there are feedback paths and many flip-flops or memory elements in the circuit. These memory elements hold a value at their outputs. For the sake of argument, even if we assume that we can apply all test vectors by observing the response at the output pins, we cannot make deterministic conclusions because of the intermingling of the values held in the feedback paths and memory elements. As will be shown in Chapters 8, 9, and 10, special designs are required to ensure the testability of a sequential circuit. This difficulty in testing is popularly known as the problem of *observability and controllability* of the internal nodes. We will further elaborate this in the next section.

Besides the technical problems, there is another consideration, which is related to human nature. In general, a person prefers to work within a narrow definition of responsibilities. A design engineer considers fabrication, testing, and quality control beyond his or her responsibilities and the practitioners of those areas of expertise feel likewise. Thus, the integration of testing and design becomes difficult.

Finally, from an economic viewpoint, a project manager must decide how much time should be spent on testing. While a higher testing cost may result in less profit, poor testing may cost the company's reputation and a loss of market. As discussed above, exhaustive testing is not feasible. Hence, the possibility of bad chips exists in every shipment. If the yield of the manufacturing process is Y, and fault coverage is T under the assumed fault model, the possibility of accepting a bad chip (defect level), DL is given by [1]:

$$DL = 1 - Y^{(1-T)} \tag{1.1}$$

This expression shows that the escape rate of bad chips may become intolerably high if a high fault coverage has not been achieved during testing.

Example 1. Consider a manufacturing process with 90% yield. All chips manufactured in this process are subjected to testing that provides 80% fault coverage. Therefore, the shipping defect level is about 2%. In other words, the probability that a chip in a shipment is good is 0.98. Thus, the probability of a good board with 50 chips is $0.98^{50} = 0.364$, which implies that 63.6% boards will be defective. Clearly, a company cannot afford such a high rate of defective boards.

1.3 COMPUTATIONAL COMPLEXITY OF TESTING PROBLEM

The first issue in estimating the complexity of the fault detection problem is to know whether a boolean expression can be satisfied. In other words, the question is to find an assignment of 1s and 0s for which the value of the expression is 1. In [2] and [3], this *satisfiability* problem has been shown to be NP-complete. Therefore, at our present level of knowledge, one cannot guarantee solving this problem in polynomial time. In [2–3], the fault detection problem in a general combinational circuit is also shown to be NP-complete. Even if we restrict ourselves to unnegated gates (monotone circuits) or to a circuit that has only negated or only unnegated gates (unate circuits), the fault detection problem is still NP-complete [3]. However, for reconvergent-free circuits, this problem can be solved in $O(n)$, where n is the number of lines in the circuit.

The complexity of the fault decision problem exists due to reconvergent fan-out paths in the circuit. These reconvergent fan-out paths require backtracking in the test generation process, which results in NP-completeness. As shown in [3], even if we limit the number of fan-outs to two, the fault detection problem is still NP-complete. We mention that the complexity of fault detection in a sequential circuit is much higher than the complexity of fault detection in a combinational circuit. This higher level of difficulty is due to the feedback paths and memory elements in the sequential circuits.

The basic problem in the testing of VLSI circuits arises from the small number of I/O pins with respect to the gate count. This small number of I/O pins limits our observation as well as our capability to control the value at an arbitrary node in the circuit. An internal node can neither be accessed directly, nor observed directly. A logic value can only be applied at the input pins, and then it must propagate through part of the circuit until it reaches the desired node. Similarly, a value at a node can only be observed by propagating it through the circuit until it reaches to the output pins. A node at which a logic value can be directly applied is called a *primary input*, and a node which can be observed directly is called a *primary output*.

One may assume that, as the number of primary inputs and primary outputs increases, the difficulty in observing and controlling a node decreases. However, for many circuits, this notion may not be true. Observability and controllability not only depend on the number of I/O pins, but also on the circuit topology. Also possible is that a particular node can be easily set to logic 1 (0), but cannot be set to logic 0 (1). Similarly, a logic value may be easily propagated from a node, while the complementary value may have difficulty in propagation.

The separate consideration of controllability and observability effectively divides the fault detection problem into two subproblems. Controllability determines the existence of an input pattern which sensitizes the fault, and observability determines the propagation of fault effect to the primary outputs. The computational complexity of these two subproblems directly relates to the complexity of the fault detection problem. Both controllability and observability in a general combinational circuit are polynomially transformable to the satisfiability problem, and hence are NP-complete. However, if we restrict ourselves to only unate circuits, controllability is solvable in $O(n)$. With a further restriction, only in two-level unate circuits, observability is solvable in $O(n^2)$ [3], where n is the number of lines in the circuit.

1.4 ESTIMATION OF DIFFICULTY IN CONTROLLABILITY AND OBSERVABILITY

In the last section, we discussed the general computational complexity of controllability and observability. We also mentioned that this problem is harder for sequential circuits as compared to the combinational circuits. As the level of difficulty varies from circuit to circuit, we need some kind of quantitative measure to understand this problem. By using arbitrary numbers, such an attempt has been made in [4], where six parameters are used to quantify the controllability, and the observability of a node:

1. CC^0: Combinational controllability to set a node to 0;
2. CC^1: Combinational controllability to set a node to 1;
3. CO: Combinational observability of a node;
4. SC^0: Sequential controllability to set a node to 0;
5. SC^1: Sequential controllability to set a node to 1;
6. SO: Sequential observability of a node.

A combinational node is defined as the input or the output of a combinational gate, while a sequential node is the input or the output of a flip-flop. The parameters CC^0 and CC^1 at node N measure the minimum number of combinational node assignments required to justify the value 0 or 1 at node N, respectively. The parameter CO at node N measures the number of combinational gates between node N and a primary output, and also the minimum number of combinational node assignments required to propagate the value from node N to a primary output. Note that at a primary input $CC^0 = CC^1 = 1$, while at a primary output $CO = 0$.

The sequential controllability parameters (SC^0 and SC^1), measure the minimum number of time frames required to control that node. The parameter SO at node N measures the minimum number of sequential cells between node N and a primary output and also the number of sequential cells which need to be controlled to propagate the value from node N to the primary output. At a primary input, $SC^0 = SC^1 = 1$, while at a primary output $SO = 0$.

The algorithm to calculate the controllability and observability for a circuit is given as follows:

1. Initialize the node controllabilities by setting $CC^0 = CC^1 = 1$, and $SC^0 = SC^1 = 0$ for the primary inputs. For all other nodes, set $CC^0 = CC^1 = SC^0 = SC^1 = \infty$.
2. Starting from the primary inputs to the primary outputs, calculate cell output node controllability by using cell input node controllabilities. Iterate this step until the controllability number stabilizes.
3. Initialize $CO = SO = 0$ for the primary outputs and for all other nodes $CO = SO = \infty$.
4. Starting from the primary outputs to the primary inputs, calculate cell input node observability by using output node observability and node controllability. Iterate this step until node observability number stabilizes.

The procedure to estimate the controllability and observability at an internal node is best illustrated with an example.

Example 2. Consider the circuit given in Figure 1.1. First, the controllability and observability values are calculated for each node. This can be illustrated by considering the OR gate. The 0 controllabilities at node Y are given as

$$CC^0(Y) = CC^0(X_1) + CC^0(X_2) + 1$$

$$SC^0(Y) = SC^0(X_1) + SC^0(X_2)$$

Similarly, the 1 controllabilities at node Y are given as

$$CC^1(Y) = \min[CC^1(X_1), CC^1(X_2)] + 1$$

$$SC^1(Y) = \min[SC^1(X_1), SC^1(X_2)]$$

The observability equations for node Y are

$$CO(X_i) = CO(Y) + CC^0(X_{3-i}) + 1; i \in [1,2]$$

$$SO(X_i) = SO(Y) + SC^0(X_{3-i})$$

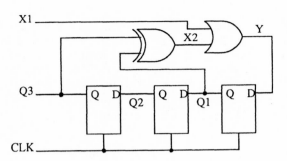

Figure 1.1 An example circuit to illustrate controllability and observability calculations. (From [4], © 1980 IEEE. Reprinted with permission.)

Each gate and flip-flop is analyzed similarly, and the controllability and observability equations are obtained. Once the expressions for every gate are obtained, using the above algorithm, the controllability and observability of each node in the circuit are calculated. For our example circuit, these values are given as follows:

Node Name	CC^0	CC^1	CO	SC^0	SC^1	SO
CLK	1	1	25	0	0	8
Y	15	2	6	4	0	3
X_2	13	26	8	4	8	3
X_1	1	1	20	0	0	7
Q1	17	4	4	5	1	2
Q2	19	6	2	6	2	1
Q3	21	8	0	7	3	0

These values are sorted and plotted in Figure 1.2. From Figure 1.2(a), it is clear that in this circuit, it is difficult to set nodes to 0 compare to 1. Also, the CC^1 for node Y is the highest, implying that it is most difficult to set a 1 at Y. The sequential controllability profiles shown in Figure 1.2(b), also indicate that 0-controllability is harder for this circuit. The CO and SO profiles are shown in Figure 1.2(c) and 1.2(d), respectively.

Although this analysis of controllability and observability provides some arbitrary numbers, this analysis is very useful in the consideration of testing strategy. These testability measures can be used effectively to insert a test point in the circuit. In automated synthesis, by using these testability measures, one may select a more testable circuit topology. Finally, in test generation (particularly in random testing), one may bias a test generator to improve its efficiency.

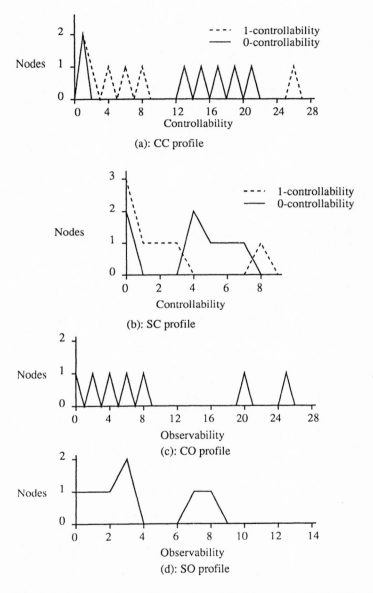

Figure 1.2 Controllability and observability profiles for Example 2. (From [4], © 1980 IEEE. Reprinted with permission.)

1.5 SUMMARY

In this chapter, we have discussed the importance of testing and various factors in the testing considerations. The computational complexity of the testing problem in the combinational circuits is discussed as NP-complete. This difficulty attributed to the difficulty in controllability and observability. For a general circuit, both controllability and observability are NP-complete problems. Finally, to quantify the testability, we discussed a procedure to quantify controllability and observability.

PROBLEMS

1. The yield of manufacturing process of a SRAM vendor is 95%. In this company, the ICs are tested for 95% fault coverage. What is the defect level? A designer used 100 chips from this company to make a memory bank. What is the probability that this board is bad?
2. Verify the controllability and observability numbers as calculated in Example 2.

REFERENCES

1. T.W. Williams and N.C. Brown, "Defect Level as a Function of Fault Coverage," *IEEE Trans. Comp.*, **30**(12), pp. 987–988, December 1981.
2. O.H. Ibarra and S.K. Sahni, "Polynomially Complete Fault Detection Problems," *IEEE Trans. Comp.*, **24**(3), pp. 242–249, March 1975.
3. H. Fujiwara, "Computational Complexity of Controllability/Observability Problems for Combinational Circuits," *IEEE Trans Comp.*, **39**(6), pp. 762–767, June 1990.
4. L.H. Goldstein and E.L. Thigpen, "SCOPE: Sandia Controllability/Observability Analysis Program," *Proc. Design Auto. Conf.*, pp. 190–196, 1980.

Chapter 2
Faults in Digital Circuits

2.1 INTRODUCTION

Many physical defects occur during the manufacturing of any system. In the manufacturing of VLSI circuits, physical defects may occur during numerous physical, chemical, and thermal processes. A defect may occur during the silicon crystal formation as well as during oxidation, diffusion, photolithography, metallization, and packaging. Some of these defects occur in the active circuit area, while others occur in the inactive area. Not all defects that occur in the active area affect the operation of the circuit. Defects that do not affect the circuit behavior only need to be examined from the reliability point of view. From the circuit operation point of view, these defects can be assumed irrelevant. The discussions in this book are primarily limited to those defects which manifest themselves as a logical fault, thus affecting the circuit operation.

Failure is generally defined in terms of circuit operation. A system failure is recognized if a system does not perform the job that it is supposed to do. This failure occurs because of some error in the operation. The error is a manifestation of a fault, while the fault is a manifestation of a defect. All defects do not result in faults; similarly, all faults do not cause an error or result in a failure. Although such testing may not be feasible, the objective is to detect all defects which affect the circuit behavior in any respect. To facilitate the testing, the assumed fault model may be directly correlated to the physical defects, or it just may be the functional description of the circuit. In the past, gate-level fault models were extensively used. In recent years, research effort has been directed to cover more realistic faults based on the analysis of physical defects. In this chapter, we describe various fault models applicable to VLSI circuits.

2.2 GENERAL VLSI FAULT MODELS

A fault model is a representative description of errors or the effect of defects. The most popular VLSI fault models are developed to cover the logical errors in the circuit operation. The actual fault coverage by these models may not be very good, but they are extremely convenient to use. Recently, to overcome the deficiency of these models, more comprehensive fault models have been developed. The computational complexity of the test generation process increases significantly with these new fault models. Thus, the logical fault models are still widely used.

2.2.1 Stuck-at Fault Model

The earliest and simplest fault model is to consider that any line in a circuit may have a fault which causes it to remain permanently either at logic 1 or at logic 0. If the logic value at a line remains at 1, the line is called *stuck-at-1* (s-a-1) and if the value remains at 0, the line is called as *stuck-at-0* (s-a-0). Any line in a circuit may have any stuck-at fault. Hence, two faults are possible on any given line. A circuit may have only one line stuck-at fault at a time, or it may have many lines with stuck-at faults. The assumption that at a given time only one line is faulty, popularly known as the *single fault assumption*. When there are many faulty lines in the circuit, the fault is called a *multiple fault*. The experience in testing dictates that, if a test set provides a good coverage of all single stuck-at faults, it also provides a good coverage of multiple faults. The coverage of multiple faults by single fault test set is not guaranteed and there may be many multiple faults in a circuit which are not covered by the single fault test set.

To test a stuck-at fault on a line, it should be sensitized by the logic value opposite to the stuck-at value. For example, to test a s-a-1 fault on a line, a 0 should be applied to this line. If this is an internal line in the circuit, the primary inputs should be controlled such that, in a fault free circuit, this line will have logic 0. After sensitizing the fault, the fault effect should be observable at a primary output. If no input combination exists at the primary inputs which sensitizes the fault, or the fault effect cannot be propagated to a primary output, the fault is called *redundant*. This situation tells us that there may be a physical defect which causes a stuck-at fault on the line. However, such a fault will not change the circuit output for any input combination.

To illustrate this point, consider the circuit given in Figure 2.1(a). In this circuit, any stuck-at fault at line X_2 can be sensitized. However, the fault effect is not observable at the output F. As both s-a-1 and s-a-0 faults on line X_2 are redundant, line X_2 is redundant. Indeed, if someone examines the output, it only depends on X_1. Such redundancies in VLSI circuits exist unintentionally as well as intentionally. As we will learn in Chapter 5, these redundant faults can be identified

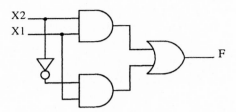

Figure 2.1(a) A sample circuit to show the effect of redundancy.

during test generation process. A redundant fault cannot be tested by logic testing. It does not affect the logic operation of the circuit, but it may affect the circuit reliability.

Apart from the redundant fault, the stuck-at fault model cannot detect many other physical defects. Consider the short circuit shown in Figure 2.1(b). This fault does not cause any electrical node to be stuck-at, rather it changes the functionality of the circuit. Obviously, the fault cannot be detected using stuck-at fault model.

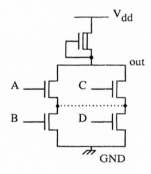

Figure 2.1(b) A sample circuit to show the inadequacy of the stuck-at fault model.

2.2.2 Bridging and Open Fault Model

The short circuit and open circuit are the two basic defect modes in any electrical circuit. Various studies since the late 1970s also suggest that the basic failure modes in VLSI circuits are physical short and open circuits [1–5]. The short circuit faults in VLSI are popularly termed *bridging faults*. These studies reported that only a small fraction of these bridging and open faults can be modeled at the stuck-at level. Many chips and layouts have been examined in these studies. The conclusions are as follows:

1. The silicon wafer defects are found in clusters. These clusters are randomly distributed over the whole wafer. Every part of the wafer has equal probability of having a defect cluster.
2. Any part of a diffusion, polysilicon, or metal line may have an open fault. Any contact between any two layers may be open.
3. A bridging may occur between any two electrical nodes, whether they belong to one layer or different layers. Bridging among multiple nodes is equally likely.
4. Only a small percentage of bridging and open faults can be modeled at the stuck-at level. The actual distribution varies and largely depends on the technology and fabrication process.

To overcome the poor fault coverage of the stuck-at fault model, a comprehensive fault model includes both bridging and the open faults. It should be noted that bridging faults are not limited to low resistance shorts (hard shorts). Any two electrical nodes may be shorted with any resistance value. The capacitive coupling (crosstalk) between two nodes is not considered to be a bridging fault. As discussed in the next section, crosstalk is considered a parametric fault.

Lately, a further division has been made by separately considering transistor stuck-on and transistor stuck-open faults. The transistor stuck-open fault is a subset of open faults. Similarly, the transistor stuck-on fault is a subset of bridging faults, a bridge between the source and drain of a transistor.

In the literature, a transistor stuck-on fault is represented in two ways: by replacing the transistor by a small resistance, or by breaking the input signal into two so that the corresponding complementary n-MOS/p-MOS will not be affected. Unfortunately, none of the approaches accurately capture the fault effect. The on-resistance of a MOSFET is in kΩ range. For a particular process, one may examine the MOS transfer characteristics to estimate the on-resistance. By replacing the MOSFET with a small resistance violates the basic voltage distribution. However, breaking the signal into two changes the line capacitances, and hence there is a loss in accuracy. This suggests that to represent a transistor stuck-on fault, one should estimate the transistor on-resistance and add that resistance between the source and the drain. For transistor stuck-open faults, breaking the signal into two and assigning different values is also inappropriate. This fault can be represented by a very high resistance (of the order of MΩ) in series with the transistor.

As bridging and open faults are the major failure modes, they need extensive and careful study. These faults are considered separately in different chapters. Detailed discussions of bridging faults are given in Chapter 3, and open faults are discussed in Chapter 4.

2.2.3 Fault Equivalence, Dominance, and Collapsing

A VLSI circuit has a large number of possible faults. If there are n lines in the circuit, there are $2n$ possible single stuck-at faults. Each transistor alone may have three bridging and three open faults. For example, Figure 2.2 illustrates 25 physical shorts and opens in a simple two-input NAND gate. This includes a majority of possible faults, but by no means includes all possible faults. Obviously, one cannot individually test such a large number of faults. To facilitate testing, the concept of fault equivalence and dominance is used to collapse the total number of faults into a small set.

   Consider the NAND gate of Figure 2.2. This gate has three lines, A, B, and Z. As each line may be s-a-0 or s-a-1, there are six possible single stuck-at faults. However, if we examine the fault effect, any s-a-0 fault at an input will result in an s-a-1 fault at the output. Therefore, an input s-a-0 cannot be distinguished from the output s-a-1 fault. In other words, these faults are equivalent.

   Now consider the line A s-a-1 fault in Figure 2.2. This fault can be detected by setting $A = 0$ and $B = 1$. If the fault is not present, the output Z is 1, while in the presence of fault, $Z = 0$. This means that $A = 0$ and $B = 1$ also detect

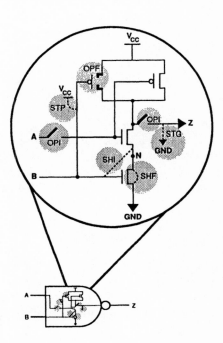

Defect Type	Number of Defects
Short-to-Ground	4
Short-to-Power	4
Open FET	4
Shorted Interconnect	4
Shorted FET	3
Open Interconnect	6
	25

Figure 2.2 Various faults in a CMOS NAND gate. Courtesy of CrossCheck, Inc.

fault line Z s-a-0. In other words, fault line A s-a-1 dominates fault line Z s-a-0. Similarly, B s-a-1 dominates Z s-a-0.

Fault equivalence and dominance allow us to combine many faults into a single set and a single test vector can detect these faults. The process of reducing the total number of possible faults into a minimal number of necessary faults is called fault collapsing. Fault collapsing is extremely important for VLSI circuits from a practical point of view. Although fault equivalence may also exist in different parts of the circuit, during fault collapsing, only the fault equivalence and dominance around individual gates are considered.

2.2.4 Parametric and Transient Faults

Apart from the bridging, open, and stuck-at faults, a VLSI circuit may have faults that do not affect the logical behavior, but degrade the performance and reliability of the circuit. These are called *parametric faults*. These faults are the major potential reliability problem. At any future instant, these faults may result in a logical error. Parametric faults include the substrate leakage current and gate oxide leakage current, variation in the threshold voltage, capacitive coupling, or crosstalk. Excessive delay in the circuit is also a parametric fault. However, in recent years, this fault has been separately considered. Delay faults are discussed in the next subsection. The cause of a parametric fault may be a physical defect or the variation in a process parameter. As these faults mainly affect the reliability of the circuit, detailed discussions and testing methods for them are given in Chapter 12.

Another possibility is the intermittent or transient fault. These faults affect the circuit behavior at random with respect to time. The cause of such faults again may be a physical defect or an environmental factor. External electromagnetic interference and ionization radiation have been repeatedly reported as a cause of transient faults. Single-bit upset and soft errors in dynamic memories due to ionization radiation are well known phenomena. A detailed discussion on transient faults is beyond the scope of this book. Interested readers are referred to [6–9].

2.2.5 Delay Fault Models

In general, from circuit specifications, whole testing problems can be divided into two categories: (1) verify the input-output logic relations and (2) ensure that the output timings in the manufactured circuit meet the design specifications. To test a timing fault, two popular models are used:

1. Single-gate delay fault model;
2. A path-oriented delay fault model.

Gate-oriented delay fault models have been used in the past [10–11]. As only a few different kinds of gates are used in a combinational block, the test length is small if the gate delay fault model is used. However, the major disadvantage is that such models cannot capture the cumulative effect of small delay variations from the primary inputs to the primary outputs. Although, a single gate may satisfy the timing specifications, the circuit may still malfunction because of the cumulative effect of delay variations.

Path-oriented delay fault models overcome this disadvantage. In path delay fault models, the cumulative delay of various gates in a path is considered as a test parameter. The transitions from 0-to-1 and 1-to-0 are propagated along a path from the primary inputs to the primary outputs [12]. A fault is detected if any of the transitions do not satisfy the specified range of time. In this procedure, a set of paths is selected such that at least one path in the set will exhibit maximum modeled delay. In an arbitrary circuit, there are a large number of possible paths from the inputs to the outputs. The size of one path may significantly differ from the others. Generally, the delays in the longest and the shortest paths (called critical paths) are examined. If these delays reside within the clock width, the circuit is considered fault-free; otherwise, the circuit has a delay fault.

2.3 SPECIFIC FAULT MODELS

In the last section, we discussed some generalized fault models. For a particular circuit, there may be some situations that cannot be described by these fault models. In this section, we consider three specific circuits. PLAs, RAMs, and microprocessors, and we discuss the fault models to develop efficient test methods.

2.3.1 PLA Fault Model

Generally, a PLA fault model consists of line stuck-at faults, bridging faults, and missing or extra crosspoint faults. In some PLAs, each crosspoint can be programmed to be on or off; hence, for those PLAs, programmability of a crosspoint switch needs to be considered instead of an extra or missing crosspoint fault. A generalized PLA fault model includes the following:

1. *Line Stuck-at Faults*: All single and multiple line stuck-at faults are considered under this category. This includes stuck-at fault at the input lines, product lines, and output lines. All stuck-at faults in the input and output registers or the pull-up logic also are included.
2. *Bridging Faults*: All single and multiple bridging faults among the input lines, product lines, and output lines are considered here. The bridging faults among the input lines and product lines or product lines and output lines also are

included. Bridging faults among the input and output lines also are possible. However, the possibility of such a fault is small. Further, such faults create feedback paths in the circuit. Unless very high fault coverage is required, these faults are not considered during testing.

3. *Crosspoint Faults*: A crosspoint may exist at an unwanted location, or a crosspoint may be missing from an intended location. These faults are known as *extra and missing crosspoint faults*, respectively. In the literature, these faults are also known as crosspoint *growth and shrinkage faults*. In some PLAs, a switch is used at the crosspoint, which can be turned on as well as off. Because the state of the switch defines whether a crosspoint is present, testing for programmability covers any crosspoint fault. To be consistent, we shall use the term extra and missing crosspoint faults to represent a switch stuck-on and stuck-off, respectively.

A multiple fault in a PLA may consist of more than one fault of one category, or more than one fault of two or more categories. All the logical faults in a PLA can be included in the above three categories. In PLAs, the line open faults are equivalent to s-a-0 faults and transistor stuck-open faults are equivalent to missing crosspoint faults. Hence, any open fault need not be separately considered. In addition to these faults, a PLA also may have parametric or timing faults.

2.3.2 Memory Fault Model

A widely used fault model for RAM devices is the one given by Nair, Thatte, and Abraham [13]. In this model, a circuit is divided into three blocks, i.e., memory cell array, decoder circuit, and the sense amplifier or the read-write circuit. In a memory array, a cell may have a stuck-at-1/0 fault or a cell may have a coupling fault with any other cell. In the decoder circuit, a decoder may not access the addressed cell, it may access a nonaddressed cell or multiple cells. The read-write circuit may have stuck-at-1/0 faults, which appear as memory cell stuck-at faults. Actual fault mechanisms based upon physical defects in memory devices have been investigated [14]. All the faults in a memory could be covered by the fault model given by Nair *et al.* with the addition of state transition faults and data retention faults. A more general fault model for memories would thus include the following:

1. Memory cell stuck-at-1/0 faults;
2. Memory cell state transition 1-to-0 and 0-to-1 faults;
3. Memory cell bridging faults with other cells (state coupling);
4. Stuck-at, multiple access or wrong addressing faults in the decoder;
5. Data retention faults.

For memories, another fault model has been widely used, particularly in the academic community. This fault model is popularly known as the *pattern-sensitive fault* model [15]. According to this model, in the presence of some specific data in a part of memory (location *i*), the data in some other part (location *j*) may be affected. This effect may appear as a cell stuck-at fault or cell state transition fault for a small duration or as long as location *i* contains the specific data. The reason for such an influence is reported as a special case of state coupling. This fault model requires that an *n*-bit memory should be tested as an *n*-state finite-state machine. Obviously, such testing is not feasible in practice. To overcome this difficulty, a simplified method is used, which is known as *neighborhood cell pattern-sensitive fault model*. In this simplified model, only the data in the neighborhood cell (physical neighbor or the logical neighbor according to the cell address) may affect the state of a cell.

In general, a memory also may have a small number of faults which can only be modeled as transistor stuck-on or stuck-open faults. Apart from these faults, a memory also may have parametric and timing or delay faults. Note that the data retention faults are a subset of timing faults. The timing faults also include data access time, which is a very important parameter for memories.

2.3.3 Microprocessor Fault Model

Microprocessors are the most complex circuits. In general, the controllability and observability in the control section of a microprocessor is extremely poor. Thus, the testing for stuck-at, bridging, and open faults is extremely difficult and time consuming, and still does not provide adequate fault coverage. To overcome this difficulty, fault models have been developed to verify the functionality of a microprocessor [16–17]. This functionality is defined in terms of the microprocessor instruction set. In [17], one assumes that each instruction is made up of multiple microinstructions, while each microinstruction is made up of multiple micro-orders. The presence of a fault is assumed under the following situations:

1. One or more micro-orders of an instruction do not execute, which means that the instruction is not executed completely.
2. In addition to the specified micro-orders, some extra micro-orders are executed with an instruction.
3. For an instruction, some specified micro-orders do not execute; instead, some other unspecified micro-orders are executed.

Note that this fault model does not cover any physical defect. A microprocessor tested under this model may have many physical defects directly related to bridging, open, or stuck-at faults. Also, timing and parametric faults are not included here. Instruction execution speed of a microprocessor is an important parameter, which should not be neglected during testing.

2.4 SUMMARY

In this chapter, we have discussed various fault models applicable to digital VLSI circuits. For logic testing, the stuck-at fault model is the simplest and most widely used. Although the stuck-at fault model does not provide adequate coverage of realistic faults, it is extremely convenient to use. Bridging and open fault models are more realistic, but they require a very elaborate test method. We also discussed specific fault models for PLAs, RAMs, and microprocessors. The logic testing of these devices can be done much more efficiently by using functional fault models.

In addition to fault models for logic testing, we also discussed fault models for reliability and parametric testing. In recent years, this topic has gained much attention. To compete in the market an extensive reliability testing is widely recognized to be necessary.

PROBLEMS

1. Consider the circuit given in Figure 2.1(a). Assume that the noninverting gates have been made by using an inverter at the output of the NAND and the NOR gates. Make a table for all possible stuck-at, bridging, and open faults for this circuit.
2. Identify all possible stuck-at faults in the circuit of Figure 2.3. Collapse this fault set into minimally necessary faults.

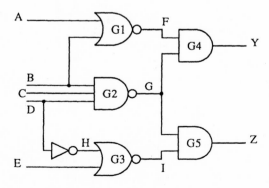

Figure 2.3 A combinational circuit for Problem 2.

REFERENCES

1.   E.I. Muehldorf, "A Quality Measure for LSI Components," *IEEE J. Solid State Cir.*, **9**(5), pp. 291–297, Oct. 1974.
2.   G.R. Case, "Analysis of Actual Fault Mechanisms in CMOS Logic Gates," *Proc. Design Auto. Conf.*, pp. 265–270, 1976.
3.   J. Galiay, Y. Crouzet, and M. Vergniault, "Physical versus Logical Fault Models in MOS LSI Circuits, Impact on Their Testability," *Proc. Int. Symp. Fault Tol. Comp.*, pp. 195–202, 1979.
4.   F. Fantini and C. Morandi, "Failure Modes and Mechanisms for VLSI ICs—A Review," *IEE Proc.*, Part G, **132**(3), pp. 74–81, June 1985.
5.   W. Maly, "Modeling of Lithography Related Yield Losses for CAD of VLSI Circuits," *IEEE Trans. CAD*, **4**(3), pp. 166–177, July 1985.
6.   T.C. May, "Soft Errors in VLSI: Present and Future," *IEEE Trans. CHMT*, **2**(4), pp. 377–387, Dec. 1979.
7.   O. Tasar and V. Tasar, "A Study of Intermittent Faults in Digital Computers," *Proc. AFIPS Conf.*, pp. 807–811, 1977.
8.   M.A. Breuer, "Testing of Intermittent Faults in Digital Circuits," *IEEE Trans. Comp.*, **13**(3), pp. 241–245, March 1973.
9.   S.Y.H. Su, I. Koren, and Y.K. Malaiya, "Diagnosis of Intermittent Faults in Combinational Networks," *IEEE Trans. Comp.*, **18**(6), pp. 567–569, June 1978.
10. D. Brand and V.S. Iyengar, "Timing Analysis Using Functional Analysis," *IEEE Trans. Comp.*, **37**(10), pp. 1309–1314, Oct. 1988.
11. Y.K. Malaiya and R. Narayanswamy, "Testing for Timing Failures in Synchronous Sequential Integrated Circuits," *Proc. Int. Test Conf.*, pp. 560–571, 1983.
12. V.S. Iyengar, B.K. Rosen, and J.A. Waicukauski, "On Computing the Sizes of Detected Delay Faults," *IEEE Trans. CAD*, **9**(3), pp. 299–312, March 1990.
13. R. Nair, S.M. Thatte, and J.A. Abraham, "Efficient Algorithms for Testing Semiconductor Random Access Memories." *IEEE Trans. Comp.*, **27**(6), pp. 572–576, June 1978.
14. R. Dekker, F. Beenker, and L. Thijissen, "A Realistic Fault Model and Test Algorithm for Static Random Access Memories," *IEEE Trans. CAD*, **9**(6), pp. 567–572, June 1990.
15. J.P. Hayes, "Detection of Pattern Sensitive Faults in Random Access Memories," *IEEE Trans. Comp.*, **24**(2), pp. 150–157, Feb. 1975.
16. S.M. Thatte and J.A. Abraham, "Test Generation for Microprocessors," *IEEE Trans. Comp.*, **29**(6), pp. 429–441, June 1980.
17. D. Brahme and J.A. Abraham, "Functional Testing of Microprocessors," *IEEE Trans. Comp.*, **33**(6), pp. 475–485, June 1984.

Chapter 3
Bridging Faults in Random Logic

3.1 INTRODUCTION

Bridging faults have long been regarded as a major failure mode in digital systems. With shrinking geometries, the study of bridging faults has become more important. A recent study, based on layout level defects, using statistical data from the fabrication process, suggests that bridging faults can be up to 40% of all faults [1].

Most of the publications on bridging faults assume a simple gate-level model for the faults. In this model, a pair of bridged lines effectively have either an AND or an OR product of the normal logic values. This model assumes the bridge to be a hard short. The resistance of the bridge has not been taken into account. In this chapter, realistic conditions are considered to evaluate the effect of different types of bridging faults. No assumption is made regarding the resistance of the bridge. Static and dynamic combinational circuits are examined for different faults.

3.2 CHARACTERIZATION OF BRIDGING FAULTS

For systematically studying the effect of bridging faults, they are characterized into three classes:

1. *Bridging within a logic element without feedback.* A logic element (LE) is a primitive gate or a complex gate. Within an LE, all internal nodes cannot be characterized by Boolean 1 or 0. The study of a bridge within a LE can be done only at the transistor level. Depending upon the resistance, the overall effect of bridging between two internal nodes may be a logical fault.
2. *Bridging between two LEs without feedback.* In this class of bridging faults, only bridging of logical nodes is considered. If the logic value of one node depends on the other node, the bridging fault is considered to be of the next

class. While the nodes being bridged are logical nodes, a transistor level examination is necessary to evaluate their effect.

3. *Bridging faults between logical nodes with feedback.* If the logic value of one node depends on the logic value of the other node, analysis becomes complex because of the feedback. This situation can lead to an unstable value at the bridged nodes. As discussed later, this may cause oscillations in the circuit as well as lead to a metastable state.

Examples of three kinds of bridging faults are given in Figure 3.1. A bridging fault between internal nodes of two LEs is possible in VLSI circuits. Test procedures to detect such faults may become quite complex. Also, bridging is possible between internal nodes in the *n*-part and the *p*-part. However, as n-MOS and p-MOS transistors are fabricated in separate wells, the probability of such faults is extremely small.

3.3 BRIDGING WITHIN A LOGIC ELEMENT

A MOSFET in a digital circuit is characterized as a switch, and hence functionally modeled as a three-terminal device (neglecting the substrate connection). A large number of bridging faults are possible within an LE. The most likely faults are characterized as follows:

1. Gate-to-drain bridging;
2. Gate-to-source bridging;
3. Source-to-drain bridging.

A three input NAND gate is examined for bridging between internal nodes. The n-MOS circuit is given in Figure 3.2(a) and CMOS circuit is given in Figure 3.2(b). The results are summarized in Table 3.1(a) and Table 3.1(b), respectively.

In n-MOS circuits, for faults 1, 2, and 3, the input vector (1,1,1) causes the output to be indeterminate. SPICE simulation shows voltages that would be recognized as logic 1 (H). Faults 4, 5, and 6 cause the V_{GS} to be less than the threshold voltage, and thus the corresponding transistor is stuck-open. Faults 7, 8, and 9 effectively cause the corresponding transistor to be stuck-on. At the gate level, these faults appear as the corresponding input s-a-1.

An interesting question is to see if the faults 1, 2, and 3 can be modeled at the gate level. A model is given in Figure 3.3. The feedback bridging looks like AND bridging, but it is more complex because of the built-in priorities of the series network. The model fails to specify the behavior for input vector (1,1,1), which is also the case with switch-level model.

In CMOS circuit (Figure 3.2(b)), faults in the *n*-part have a direct correspondence with the faults in n-MOS circuit. The following correspondence rules can be used:

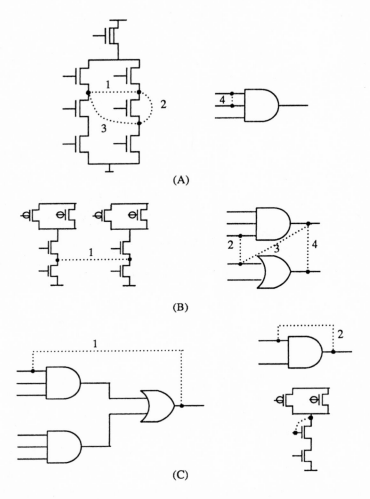

(A)

(B)

(C)

Figure 3.1 Examples of bridging faults: (a) bridging within a logic element without feedback; (b) bridging between two LEs without feedback; and (c) bridging with feedback.

1. Normal n-MOS output 0, faulty n-MOS output 0; in this case, the *n*-part retains a low resistance path from output to the GND. The *p*-part presents a high resistance path from output to V_{dd} and hence for CMOS, for the same fault, output is 0.
2. Normal n-MOS output 1, faulty n-MOS output 1; due to the same reasons, in this case for the same fault in CMOS, the output is 1.

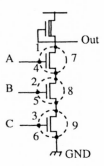

Figure 3.2(a) Bridging faults within an n-MOS NAND gate.

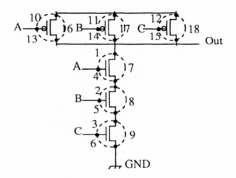

Figure 3.2(b) CMOS NAND gate with bridging faults.

3. Normal n-MOS output 1, faulty n-MOS output 0; for CMOS, in this case, the output has a low resistance path to the GND because of the fault. If the inputs are not affected by the bridging, there also is a low resistance path from the output to the V_{dd}. Thus, the output is undefined for CMOS.
4. Normal n-MOS output 0, faulty n-MOS output 1; due to the same reasons, in this case for the same fault in CMOS, the output is either undefined or in the high impedance state.

The above correspondence rules suggest that a fault which is testable in n-MOS may become untestable in CMOS. This result can also be seen in Table 3.1(b). The effect of drain to source bridging (faults 7–9 and 16–18) is not given in Table 3.1(b). Such faults are not deterministically testable by logic testing. The only definite way to test them is by monitoring the excessive leakage current

Table 3.1(a)

Output of n-MOS NAND gate (Figure 3.2(a)) in presence of various bridging faults
(u represents undefined logic output).

Inputs ABC	Normal	1	2	3	4, 5, 6	7	8	9
000	1	0	1	1	1	1	1	1
001	1	0	1	1	1	1	1	1
010	1	0	1	1	1	1	1	1
011	1	0	1	1	1	0	1	1
100	1	1	0	1	1	1	1	1
101	1	1	0	1	1	1	0	1
110	1	1	1	0	1	1	1	0
111	0	u	u	u	1	0	0	0

[3–4]. A test method based on measurement of the power supply current is described in Chapter 11.

In Table 3.1(b), wherever the switch-level analysis does not specify a definite 1 or 0, results from SPICE analysis also are given. Notice that only bridging faults 6, 10, 11, and 12 can be tested deterministically. A probabilistic view of the testability of the CMOS gate is not as pessimistic. SPICE analysis suggests that the vector (1,1,1) is very likely to test for faults 1 and 13. Vector (1,0,1) should test for fault 14 and vector (1,1,0) should test fault 15. Fault 2 appears to be the least testable.

A bridging fault that is not testable is still extremely undesirable. It will affect the dynamic behavior, reduce the noise margin, and cause the chip to age quite fast by drawing and dissipating very large currents. One way to overcome the lack of deterministic testability is to use testing methods based on monitoring supply current as described in Chapter 11. In general, a bridging fault in the CMOS gate is always detectable by current testing. Another possibility is deliberately introducing asymmetry, causing either the *n*-part or *p*-part to dominate. A third possibility is to use the reduced noise margin caused by untestable faults. Adding a controlled amount of noise will trigger some of the faults.

3.4 BRIDGING OF LOGICAL NODES WITHOUT FEEDBACK

Bridging between two nodes can represent any resistance value. The effect of bridging cannot be modeled accurately without taking into account the resistivities of the paths and resistance of the bridging. The ideal switch-level analysis cannot be used to analyze the circuit if the bridging resistance is taken into account. If

Table 3.1(b)

Output of CMOS NAND gate (Figure 3.2(b)) in presence of bridging faults.

Inputs ABC	Normal	1	2	3	4	5	6	10	11	12	13	14	15
000	1	u,1	1	1	1	1	1	1	1	1	u,1	u,1	u,1
001	1	u,1*	1	1	1	1	1	1	1	1	u,1*	u,1*	1
010	1	u,1*	1	1	1	1	1	1	1	1	u,1*	1	u,1*
011	1	u,0	1	1	1	1	1	(0),0	1	1	u,0	1	1
100	1	1	u,1	1	1	1	1	1	1	1	1	u,1*	u,1*
101	1	1	u,1*	1	1	1	1	1	(0),0	1	1	u,0	1
110	1	1	1	u,1	u1,1	u1,1	(1)	1	1	(0),0	1	1	u,0
111	0	u,1*	u,1*	u,1	u1,1	u1,1	(1)	0	0	0	u,1	u,1	u,1

u: undefined;

u1: undefined but most likely 1;

(1): H if bridge is a hard short;

(0): L if bridge is a hard short;

1*: probable H, i.e., 3.5 > voltage > 2.5;

x,y: result of switch level and circuit level analysis, respectively.

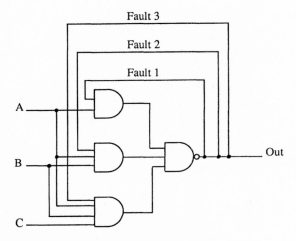

Figure 3.3 A gate-level model for bridging faults 1, 2, and 3.

the resistance is sufficiently close to zero, it is referred as a *hard short*. In the literature, a hard short has always been assumed for bridging. However, as the bridging is not a deliberate feature, it might not be well defined and generally may exhibit significant resistance. A very high resistance, approaching to infinity, implies a bridging fault of no consequence.

When two logical nodes having the same logic value are bridged together, the bridge will not have any significant effect. All cases where the nodes have opposite polarities are considered below, as shown in Figure 3.4(a). Outputs of

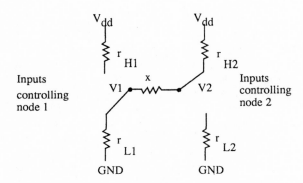

Figure 3.4(a) Computational model for bridging faults. (From [2], © 1986 IEEE. Reprinted with permission.)

two LEs, indicated by subscripts 1 and 2, are bridged by a resistance x. Let the resistances that connect the L(H) node to the ground (V_{dd}) be $r_L(r_H)$.

In order to analyze the effect of bridging, the following parameters must be considered, as shown in Figure 3.4(b):

Worst-case (min.) output H voltage $= H_w$

H-level noise-margin $= N_H$

Worst-case (max.) output L voltage $= L_w$

L-level noise-margin $= N_L$

Switching threshold voltage $= S_T$

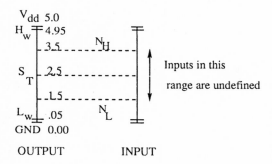

Figure 3.4(b) Assured voltage levels and noise margins. (From [2], © 1986 IEEE. Reprinted with permission.)

For an ideal switch, the switching threshold voltage is the input voltage at which the output voltage switches from one binary state to another. For an inverter, this is the input voltage for which the output and input voltages are the same. The noise margins specified in data books are generally statistical 3σ values. Noise swings less than the noise margins are very unlikely to cause the wrong level to be recognized. Some texts define noise margins with respect to 45° slope points on the voltage transfer characteristics, which also results in similar values.

Let V_1 and V_2 be the voltages at the output nodes in the presence of bridging. Then,

$$V_1 = V_{dd} \frac{r_{L1}}{r_{L1} + x + r_{H2}} \tag{3.1a}$$

and

$$V_2 = V_{dd} \frac{r_{L1} + x}{r_{L1} + x + r_{H2}}$$

(3.1b)

The bridging fault is not consequential if $x \to \infty$, giving $V_1 = 0$ and $V_2 = V_{dd}$. For a hard short, $x \to 0$, which gives $V_1 = V_2 = V_{dd} \cdot [r_{L1}/(r_{L1} + r_{H2})]$. In this analysis, r_H represents the resistance of the pull-up transistor and r_L depends on the on-resistance of the pull-down transistors. While the actual values would vary, depending upon the transistor size and processing parameters, for n-MOS, r_H is always significantly higher than r_L. Therefore, a hard short in n-MOS will always cause both nodes to be in the L range provided that the driving strengths of the nodes are the same. Hence, at logic level, a hard short is modeled as an AND bridging.

In CMOS circuits, the assumption that r_H is significantly higher than r_L, may not be valid. If the transistors are scaled appropriately to neutralize the difference in mobility of electrons and holes, r_H and r_L become comparable. Therefore, a hard short will cause an indeterminate logic level. However, if the transistor sizes are the same, because of the mobility difference, r_H is about three times larger than r_L. The ratio of r_H to r_L can be further enhanced without much degradation in noise margins. In this situation a hard short will cause both nodes to be in the L range. Hence, such fault can be modeled as AND bridging.

When the bridging has some finite resistance, the situation is more complex. Depending on the bridge resistance and the voltage strength at the bridged nodes, one of the following results can happen:

1. The bridge could be very mild, i.e., of very high resistance. There will be no effect at the logic level, but degradation in noise immunity may occur.
2. The voltage at the node may be outside of the guaranteed noise margin, but still on the correct side of the switch threshold point. The correct logic value is likely, but is not guaranteed. If there is a fan-out at the shorted node, different branches possibly may appear to have different logic values, depending upon the exact thresholds of the succeeding logic.
3. The voltage may be on the other side of the switch threshold point, but not within the noise margin of the opposite logic value. The logic value is likely to be recognized as faulty. Different branches of a fan-out at the bridged node possibly may appear to have different logic values.
4. If it is a hard short, the opposite logic value may prevail. This situation is modeled by AND bridging.

The four ranges for the values of x are defined by x_1, x_2, and x_3 in Figure 3.5. Algebraic expressions for these are obtained below. V_1 and V_2 must be considered

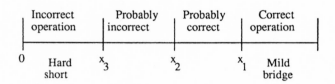

Figure 3.5 The four ranges of bridging resistance. (From [2], © 1986 IEEE. Reprinted with permission.)

separately because they generally will be different. Let us assume that V_1 is normally L and V_2 is normally H, as shown in Figure 3.4(a).

Case a. For V_1 to be normal, even in the presence of bridging:

$$V_1 \leq (L_W + N_L)$$

i.e.,

$$V_{dd} \frac{r_{L1}}{r_{L1} + x + r_{H2}} \leq (L_W + N_L)$$

which gives

$$x \geq \left(\frac{V_{dd} \cdot r_{L1}}{L_W + N_L} \right) - r_{L1} - r_{H2}$$

Hence,

$$x_1 = \left(\frac{V_{dd} \cdot r_{L1}}{L_W + N_L} \right) - r_{L2} - r_{H2} \tag{3.2a}$$

Similar condition for V_2 is given by

$$x_1 = \frac{r_{L2} + r_{H2} - \left(\dfrac{V_{dd}}{H_W - N_H} \right) \cdot r_{L1}}{\left(\dfrac{V_{dd}}{H_W - N_H} \right) - 1} \tag{3.2b}$$

Cases b and c. The value of x_2 for V_1 is given by

$$S_T = V_2$$

i.e.,

$$x_2 = \left(\frac{V_{dd} \cdot r_{L1}}{S_T}\right) - r_{L2} - r_{H2} \tag{3.3a}$$

Similarly for V_2, it is given by

$$x_2 = \frac{r_{L1} + r_{H2} - \left(\dfrac{V_{dd}}{S_T}\right) r_{L1}}{\left(\dfrac{V_{dd}}{S_T}\right) - 1} \tag{3.3b}$$

Case d. V_1 will assume the opposite logical value H if:

$$V_1 \geq (H_W - N_H)$$

i.e.,

$$V_{dd} \frac{r_{L1}}{r_{L1} + x + r_{H2}} \geq (H_W - N_H).$$

which gives

$$x_3 = \left(\frac{V_{dd} \cdot r_{L1}}{H_W - N_H}\right) - r_{L1} - r_{H2} \tag{3.4a}$$

Similarly for V_2, x_3 is given by

$$x_3 = \frac{r_{L1} + r_{H2} - \left(\dfrac{V_{dd}}{L_W + N_L}\right) r_{L2}}{\left(\dfrac{V_{dd}}{L_W + N_L}\right) - 1} \tag{3.4b}$$

Example 1. For CMOS with 5 V supply, $V_{dd} = 5$ and $(L_W + N_L) = 1.5$. The values of r_{H2} and r_{L1} can be evaluated by using appropriate parameters for SPICE simulation. A bridging fault between the outputs of a NAND and a NOR gate has been simulated, as shown in Figure 3.6. Simulation of fault-free circuit suggests that the on-resistances of n-type and p-type transistors to be approximately 4.5 kΩ and 30 kΩ, respectively. If the input vector is HHLL, then the resistances involved

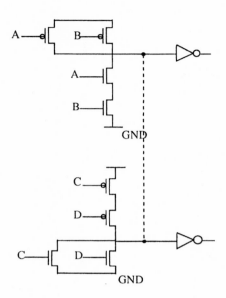

Figure 3.6 CMOS NOR and NAND gates with bridged outputs.

are: $r_{l,1} = 9$ kΩ and $r_{l/2} = 60$ kΩ. Then the values of x_1, x_2, and x_3 for V_1 are found to be -39kΩ, -51kΩ, and -16kΩ, indicating that the logic value for V_1 is always correct (L) regardless of the resistance of x. However, for V_2, the values of x_1, x_2, and x_3 are found as 131kΩ, 21kΩ, and 3.8kΩ. The simulation results confirm that the behavior changes from correct to incorrect as the resistance of bridging decreases below 20kΩ. As x_1 and x_3 both involve statistical considerations, a direct confirmation by SPICE is not possible. However, above 130kΩ, the bridge has hardly any effect and, below 4kΩ, the bridge appears as a hard short (Figure 3.7).

Some important observations can be made here. Notice that $r_{l,1}$, $r_{l/2}$, et cetera depend on transistor parameters, transistor configuration within a gate, and logical input vectors. Thus, we have the following:

1. Values x_1, x_2, and x_3 for a given bridging fault will depend on the test vector. The fault therefore will look like bridging sensitizable under only some vector. When $x_2 \le x \le x_1$, probability of sensitizing the fault will be less than 0.5 and, when $x_3 \le x \le x_2$, the probability of sensitizing the fault will be more than 0.5.
2. In some cases, bridging can have an asymmetric effect. In Figure 3.4(a), if $r_{H1} > r_{l,2}$ and $r_{l/2} < r_{l,1}$, then a L at node 2 would dominate H at node 1, but a L at node 1 would be dominated by H at node 2. This implies that a node

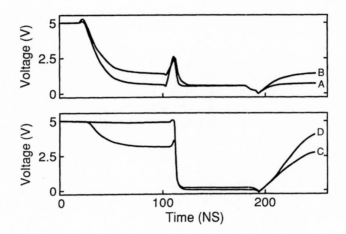

Figure 3.7 Effects of bridging on NAND output: (a) $x = 10k\Omega$, hard short; (b) $x = 10k\Omega$, $x_2 - x_3$ region; (c) $x = 50k\Omega$, $x_1 - x_2$ region; (d) $x = 1M\Omega$, mild bridge.

dominates over the other rather than one logic value dominates over the other.

3. A bridging for which $x < x_1$ is always true, will be normally undetectable. However, the reliability with respect to electromagnetic and electrostatic noise will be greatly reduced because of the reduced noise margin.

Clearly, a major effect of high resistance bridging is the reduction in noise immunity. The possibility of detecting this kind of bridging fault may be significantly enhanced in the presence of noise. Testing in an environment with a measured amount of noise would have an effect similar to accelerated testing using temperature or voltage stress as described in Chapter 12. Another possible method of detecting these faults is current monitoring. As has been reported, all single and multiple bridging faults are detectable by current testing. The details of the current testing method are given in Chapter 11.

The main limitation of the above model is the nonlinearity in transistor resistance. The model assumes a fixed resistance of the transistor. Therefore, in some cases, predictions may not be very accurate. Furthermore, the model requires an estimation of bridging resistance which might be difficult in some cases.

3.5 BRIDGING OF LOGICAL NODES WITH FEEDBACK

When one of the nodes involved in a bridging fault derives its logic value from another node, the situation becomes more complex. The behavior of a combinational block becomes sequential because of the feedback loop.

If the feedback loop does not contain clocked storage elements, asynchronous feedback paths are introduced. In general, this transforms a combinational block into an asynchronous sequential circuit. Such a circuit may exhibit stable states. If a feedback loop contains an odd number of inversions, oscillations may occur. Generally, the effect of feedback bridging has to be evaluated by using the analysis techniques for asynchronous sequential circuits.

Sometimes, feedback bridging may cause anomalous behavior, which cannot be described logically. This behavior depends not only on the propagation delay (which can be modeled at gate level), but also on the rise time and fall time as well as the analog transfer characteristics of the transistors.

Consider the situation illustrated in Figure 3.8. An input I1 of a combinational block is bridged with an output Z1 through a bridging resistance x. Assume that an input vector is applied, under which there is a sensitized path from I1 to Z1 through an odd number of inversions. Also assume that x is a small resistance and, at the node connecting I1 and Z1, L level dominates H. In the fault-free condition, when I2 is L and Z1 is H, a stable situation exists.

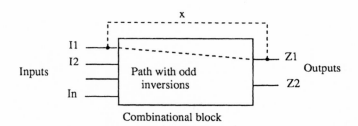

Figure 3.8 Feedback in a combinational block. (From [2], © 1986 IEEE. Reprinted with permission.)

Now consider the fault with the situation I1 is H. If the rise and fall times are negligible compared to the propagation delay, any logic value at Z1 will cause itself to change after the propagation delay. This implies that oscillations will occur, and the time period of oscillations will be equal to the propagation delay.

Another situation arises when the propagation delay is smaller than the rise or fall time. In this case, the output will start changing before the input has stabilized. This suggests that under the bridging, a falling or rising input signal will start influencing itself in the opposite direction before it reaches the switching threshold. The logic value attempts to stabilize somewhere between the H and L levels. Hence, under this condition, the fault may lead to a metastable state.

The anomalous behavior in this case cannot be analyzed by using a static analysis assuming a pure transport delay. Dynamic analysis must be performed, taking into account the capacitances associated with the nodes. Also during switching, the switch is not ideal and it will exhibit a finite and varying resistance.

To illustrate the two cases, inverter chains with a feedback bridging fault were examined using SPICE (Figure 3.9). It was found that oscillations occur when the propagation delay was sufficiently large (when the feedback path contained three or more inverters). When the propagation delay was small (one inverter), oscillations did not occur, and the voltage level stabilized at some intermediate level.

An approximate relationship using the delay parameters can be obtained which gives the condition for oscillations. An exact relation cannot be obtained because the true rise and fall times are not measurable, a 10% to 90% value range is always used.

If x (Figure 3.9) is small, either the L or the H level would dominate and the oscillations will reach either the H or the L level. When the propagation delay pd is large enough for oscillations, it will allow the signals to cross the switching threshold voltage S_T. If S_T is situated in the middle of H-to-L range, L-to-S_T and S_T-to-L transitions will take approximately half of the rise time t_r and half of the fall time t_f. The requirement for the oscillations to occur is

$$pd > \frac{t_r}{2} + \frac{t_f}{2} \tag{3.5}$$

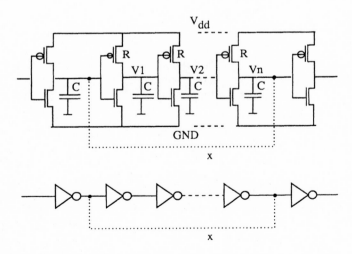

Figure 3.9 Bridging across a path with odd inversions.

If there are N inversions in the feedback path (N being odd), the condition for oscillations is [8]:

$$\left[\frac{1}{\alpha} + i\omega \cdot \frac{RC}{\alpha}\right]^N = -1 \tag{3.6}$$

where ω is the frequency of oscillations and α is the dc gain of one gate. Equation (3.6) can be rewritten as

$$\frac{1}{\alpha} + i\omega \cdot \frac{RC}{\alpha} = \cos\left(\frac{k\pi}{N}\right) + i \cdot \sin\left(\frac{k\pi}{N}\right) \tag{3.7}$$

Equating the imaginary and real parts, the amplitude and frequency of oscillations can be obtained. The amplitude of oscillations is given as

$$\frac{1}{\alpha} = \cos\left(\frac{k\pi}{N}\right) \tag{3.8}$$

When N is large, amplitude of oscillations is unity, which implies that we shall always have either the H or the L level.

For large N and small k, the frequency of oscillations is given as

$$\omega = \frac{\alpha}{RC} \cdot \left(\frac{k\pi}{N}\right) \tag{3.9}$$

When the feedback path has only one inversion ($N = 1$), the circuit is absolutely stable. However, Equation (3.5) may not be satisfied to start oscillations. In this situation, circuit shows metastability.

Example 2. SPICE simulation for a specific set of parameters for CMOS inverter chain exhibit these values: pd/gate = 7.2 ns, t_r = 7.6 ns, and t_f = 10.8 ns. This suggests that oscillations will occur if the chain has a propagation delay greater than 9.2 ns. Simulation shows that with one inverter (pd = 7.2 ns) oscillations do not occur, but the circuit oscillates with three inverters (pd = 21.6 ns). Similarly, for n-MOS, we found pd/gate = 1.6 ns, t_r = 0.7 ns, and t_f = 6 ns. Equation (3.5) gives the minimum pd required for oscillations as 3.4 ns. Indeed, for one inverter (pd = 1.6 ns), there were no oscillations (Figure 3.10), while for three inverters (pd = 4.8 ns) the circuit oscillates (Figure 3.11).

The simulation shows that if the value of resistance x is increased, the magnitude of the oscillations reduces, and one of the two levels is not always reached

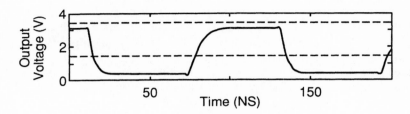

Figure 3.10 Output of CMOS inverter with feedback bridging without oscillations.

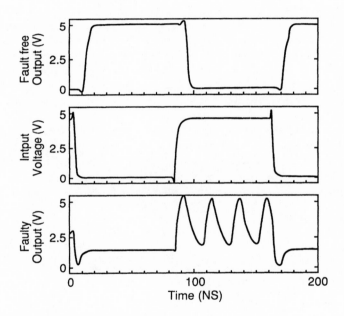

Figure 3.11 Output of circuit given in Figure 3.9.

as expected by Equation (3.8). However, the frequency of oscillations remains the same as predicted by Equation (3.9). When x is sufficiently high, the operation becomes normal as expected.

3.6 BRIDGING IN DYNAMIC GATES

In recent years, dynamic gates have received considerable attention due to their high speed performance, small area requirement, and low power consumption.

Major dynamic logic families are *domino logic*, *cascade voltage switch logic* (CVSL), and *clocked CMOS logic* (C²MOS). In this section, the effect of bridging faults in these logic families has been discussed. Other dynamic circuits like NORA and zipper CMOS are based on one of these three families [5], and hence have not been explicitly discussed.

3.6.1 CMOS Domino Logic

CMOS domino circuits are equivalent to n-MOS dynamic circuits with precharge. All the CMOS domino gates have an n-MOS driver part with p-MOS pull-up transistor, followed by a static CMOS inverter. The input node of the inverter is precharged to 1 when clock is 0. When clock is 1, called the evaluation phase, this node is conditionally discharged through the driver part. Because of the static inverter, output of the gate, and hence the inputs of the successive gates remain low during the precharge phase. As domino is a noninverting logic, the speed of operation is determined by the time the effect of 1 takes to travel through the critical path.

A two-input AND gate in domino logic is examined for possible bridging faults as shown in Figure 3.12. The effect of different bridging resistance is found the same as discussed in Section 3.3. The results in case of hard short are summarized in Table 3.2. For fault 4, 5, 6, and 10, the entries in Table 3.2 are marked P, which represent the previous value. Under these faults, corresponding transistors

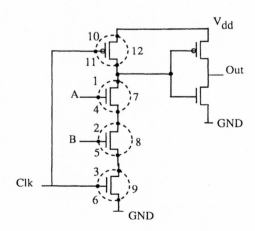

Figure 3.12 CMOS domino AND gate with bridging faults.

Table 3.2

Output of a domino AND gate in presence of various bridging faults.

Inputs ClkAB	Normal	1	2	3	4	5	6	7	8	9	10	11	12
000	1	u	1	1	1	1	1	1	1	1	p	0	1
100	1	0	1	1	p	p	p	1	1	1	p	1	1
001	1	u	1	1	1	1	1	1	1	1	p	0	1
101	1	0	1	1	p	p	p	0	1	1	p	1	1
010	1	1	u	1	1	1	1	1	1	1	p	0	1
110	1	1	0	1	p	p	p	1	0	1	p	1	1
011	1	1	1	u	1	1	1	1	1	u	p	0	1
111	0	u	u	u	p	p	p	0	0	0	0	u	u

u: undefined;

p: previous value. In this condition output node is in floating state.

are s-open, and hence the input node of static inverter is in high impedance state during the evaluation phase. These faults are discussed in Chapter 4, Section 4.5.

A two-pattern sequence is required to test a fault that creates a high impedance state at the output. The first pattern should initialize the node, while second pattern should activate the fault. In domino logic, this can be done effectively within one clock period. The precharge stage is used for initialization and the evaluation stage is used to sensitize the fault [6]. To test the fault that causes high leakage current and hence indeterminate logic output, current monitoring is recommended.

3.6.2 Cascade Voltage Switch Logic

CVSL is a differential style of logic, which requires both true and complementary signals. Two complementary n-MOS structures are connected to a pair of cross-coupled p-MOS pull-up transistors to obtain a CVSL gate. Only one n-MOS structure switches on at a given time. Positive feedback applied to the p-MOS pull-up transistors causes the gate to switch. This logic is slower than a conventional CMOS gate because during switching p-MOS pull-up transistors have to fight the n-MOS-pull-down structure.

A two-input NAND/NOR CVSL gate is examined for possible bridging faults as shown in Figure 3.13. The results in the case of a hard short are summarized in Table 3.3. Again, in Table 3.3, some entries are marked P to represent the floating output.

3.6.3 Clocked CMOS Logic

Clocked CMOS was originally developed to build low power CMOS circuits. The main use of such logic is to form clocked structures that incorporate latches or interface with other dynamic logic gates. The gates have the same input capacitance

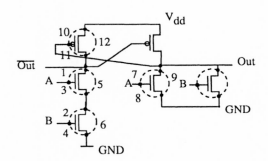

Figure 3.13 CMOS CVSL gate with bridging faults.

Table 3.3

Output of a CVSL NADN/NOR gate in the presence of various bridging faults.

Inputs AB	Normal xx'	1	2	3	4	5	6	7	8	9	10	11	12
00	10	00	10	10	10	10	10	10	10	10	pu	uu	10
01	10	00	10	10	10	00	10	p1	pp	10	pu	uu	10
10	10	10	u0	10	10	10	00	10	10	10	pu	uu	10
11	01	01	10	10	10	10	01	u0	01	u0	01	uu	10

u: undefined;

p: previous value. In this condition output node is in floation state.

as conventional CMOS gates, but have larger rise and fall times due to the series clocking transistors.

A two-input NAND gate implemented in C²MOS is examined for possible bridging faults as shown in Figure 3.14. The results in the case of a hard short are summarized in Table 3.4. When clock is 0 and input vector is (1,0), the output is in the floating state. These situations are marked P in Table 3.4. An interesting situation arises under fault 16. In the case of clock and $\overline{\text{clock}}$ not isolated, this fault causes an anomalous voltage level at the output.

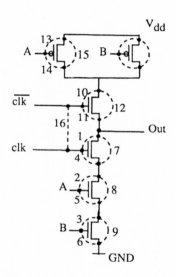

Figure 3.14 Clocked CMOS gate with bridging faults.

3.7 EFFECT OF SUBSTRATE CONNECTION

In Sections 3.3 and 3.4, simple switch-level analysis is used to model a MOSFET, neglecting the substrate connection. The results obtained generally are true because substrate connection does not play a major role. In the majority of digital ICs made on silicon, the substrate of an n-MOS transistor is connected to the ground or the maximum negative voltage available on the chip. Similarly, the substrate of a p-MOS transistor is connected to the V_{dd} or maximum positive voltage. This ensures that the substrate-drain and the substrate-source junctions never become forward-biased. Thus, the substrate (or *well*) becomes completely isolated and the possibility of current through the substrate is eliminated. In this case, the drain and source regions are interchangeable (except in cases such as *lightly doped drain*

Table 3.4

Output of clocked CMOS gate in presence of various bridging faults.

clk$\overline{clk}$ABOut	1	2	3	4	5	6	7	8	9	10	11	12	13	14	15	16
01001	0	p	p	p	p	p	p	p	p	p	1	1	p	p	p	p
10001	1	u	1	1	1	1	1	1	1	p	0	1	1	1	1	0
01011	0	1	p	p	p	p	p	p	p	p	1	1	p	p	p	p
10011	1	0	1	1	1	1	1	u	1	p	0	1	0	0	1	0
01100	0	1	p	p	p	p	p	p	p	p	1	1	p	p	p	p
10101	1	1	1	1	1	1	1	1	u	p	0	1	1	1	1	0
01111	0	1	p	p	p	p	0	p	p	p	1	1	p	p	p	p
10110	1	u	u	p	p	p	0	0	0	0	0	0	0	0	u	0

u: undefined;

p: previous value. In this condition output node is in floating state.

(LDD) type structures). Hence, the device acts as a simple switch controlled by the gate voltage.

In some IC designs, the substrate (or well) of a transistor is connected to its source node (rather than the maximum positive or negative voltages). This is generally true in *silicon-on-insulator* (SOI) technology, in which a floating substrate lowers the effective threshold voltage and thus causes a "kink" in the transfer characteristics. The main reasons for this kink include impact ionization, weak avalanche, and charge pumping in the substrate. To avoid the effects of this kink, the transistor substrate is electrically connected to its source node. Since, under normal operation of a device, the source voltage of an n-MOS (or p-MOS) transistor is lower (or higher) than the drain voltage, there is no possibility of the substrate-drain junction becoming forward-biased. In general the source-substrate connection results in lower body effect and equal potential distribution, or may even result in less area. However, when the substrate of a transistor is connected to its source, its behavior cannot always be characterized as a simple switch. Consider the following cases:

1. When drain voltage V_d of an n-MOS transistor is greater than its source voltage V_s and the gate voltage is high, the transistor conducts and provides a low impedance path between its drain and the source. When gate voltage is low, the transistor offers a high impedance between its source and the drain. The switch-level model is valid for this case.
2. When $V_d < V_s$ (which can occur only in the presence of a fault), the substrate-drain junction becomes forward-biased. This provides a low resistance source-to-drain path through the substrate and the current through the transistor is no longer controlled by the gate voltage. Therefore, the MOSFET cannot be modeled as a switch controlled by the gate. In this case, the current through the device can cause significant power dissipation, which could even result in permanent failure of the device. Such conduction of a MOS transistor is referred to as *anomalous reverse conduction* (ARC) [7]. This situation never occurs under normal operation of the device.

Bridging faults in n-MOS and CMOS complex gates were examined under both substrate connection schemes:

1. When the substrate is connected to maximum negative or positive potential;
2. When the substrate is connected to the source.

We assume that all the inputs are independently accessible and only nonredundant gates are considered. In the case where the bulk of an n-MOS or CMOS transistor is connected to the most negative (positive) voltage, the results from Switch Level Analysis (SLA) are in complete agreement with the circuit-level analysis. Hence, the results for this case are not discussed.

Consider the structure of Figure 3.15, assuming that the substrate of each transistor is connected to its source. Each bridging fault was examined under all possible test vectors at the switch level as well as at the circuit level (using SPICE). The results are shown in Table 3.5. Some faults provide only redundant information. Hence, they are not included in the table. When the value predicted by SPICE is outside of noise margins, it is termed as probably H or probably L, depending on whether it is higher or lower than the switching threshold. We can see that all the faults are deterministically testable. For most of the faults, results obtained using switch-level analysis are found to be in agreement with those from the circuit-level analysis. However, there are some cases in which the results of switch-level analysis do not match those from the circuit-level analysis. In these cases, a more detailed analysis at a lower level is essential. This is best illustrated by an example.

Example 3. Consider the n-MOS complex gate shown in Figure 3.15 in the presence of fault #10 (a short between node 5 and input D) with the test vector $ABCD = 0011$. The complex gate implements the function $\overline{AB + CD}$. A simple switch level analysis suggests that the transistor NA and NB are off, and transistors NC and ND are on. One end of the short, node 5, is between two off transistors. According to the assumptions of switch-level analysis, the presence of bridging should be of no consequence. The output, being connected to the ground through NC and ND, both of which are on, should be at logic 0.

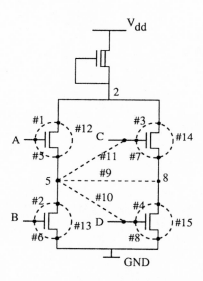

Figure 3.15 NMOS complex gate with bridging faults to show the effect of substrate connection. (From [7], © 1987 IEEE. Reprinted with permission.)

Table 3.5

Analysis of n-MOS complex gate of Example 3 for different bridging faults.

Inputs ABCD	Normal	1	2	5	6	9	10	11	12	13
0000	1	0,0	1,1	1,1	1,1	1,1	1,1	1,1	1,1	1,1
1000	1	1,1	0,0	1,1	1,1	1,1	0,0	1,1	1,1	0,0
0001	1	0,0	1,1	1,1	1,1	1,1	1,1	1,1	1,1	1,1
1001	1	1,1	0,1	1,1	1,1	1,1	0,0	1,1	1,1	0,0
0111	1	0,0	1,1	1,1	1,1	1,1	1,1	1,1	0,0	1,1
1100	0	1,1*	1,1*	1,1	1,1	0,0	0,0	0,0	0,0	0,0
0101	1	0,0	1,1	1,1	1,1	0,0	1,1	1,1	0,0	1,1
1101	0	1,1*	1,1*	1,1	1,1	0,0	0,0	0,0	0,0	0,0
0010	1	0,0	1,1	1,1	1,1	1,1	1,1	1,1	1,1	1,1
1010	1	1,1	0,0	1,1	1,1	0,0	1,1	0,0	1,1	0,0
0011	0	0,0	0,0	0,0	0,0	0,0	0,1*	0,1*	0,0	0,0
1011	0	0,0*	0,0	0,0	0,0	0,0	0,0*	1,1*	0,0	0,0
0110	1	0,0	1,1	1,1	1,1	1,1	1,1	1,1	0,0	1,1
1110	0	1,1*	1,1*	1,1	1,1	0,0	0,0	0,0	0,0	0,0
0111	0	0,0	0,0	0,0	0,0	0,0	1,1	1,1	0,0	0,0
1111	0	0,0	0,0	0,0	0,0	0,0	0,0	0,0	0,0	0,0

1*: Probable H (2.7–2.2 V);
0*: Probable L (2.2–1.25 V)
x,y: Results of switch level and circuit level analysis, respectively.

However, SPICE analysis gives an output voltage of 2 V, which is neither logic 0 nor logic 1 (see Figure 3.16). The analysis also gives a current of about 20 μA flowing from node 5 to node 2. In this case, node 5 effectively becomes the drain of NA. However, according to the switch-level model, there should be no current through NA except for a small leakage current.

The circuit level analysis suggests that the resistance of the off transistor NA is not as close to the ideal (infinite) as the switch level analysis assumes. The significantly low off resistance and thus, the abnormally large leakage current through transistor NA, can be explained by the ARC phenomenon.

Example 3 is restricted to one PCC. In this case, SLA may give misleading results for some input vectors. However, the faults are testable. When a bridging fault occurs between two nodes of two different PCCs, the analysis becomes more complicated. In such a case, the fault possibly may become untestable at switch level as illustrated in Example 4.

Example 4. Consider the n-network of the complex gate shown in Figure 3.17, which implements the function $(\overline{CD+E})\,(\overline{AB+F})$. The gate may be an n-MOS gate with a depletion-mode load transistor or a CMOS gate with a p-network as the pull-up part. Consider a bridging fault between nodes 1 and 3. Such a fault is quite possible in the given layout because the two nodes are close to each other. Consider the test vector $ABCDEF = 100010$. According to the switch level analysis, the transistors ND, NC, NB, and NF are off, and the transistors NA and NE are on. The n-network is thus off. The output should be logic 1. Also, in the case of

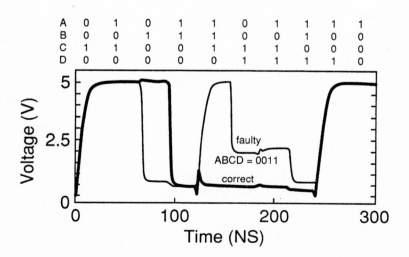

Figure 3.16 SPICE output for fault #10, in n-MOS complex gate given in Figure 3.15.

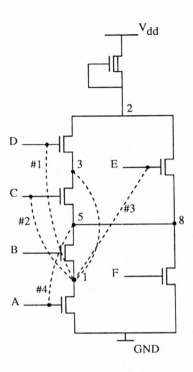

Figure 3.17 n-MOS complex gate with bridging faults for Example 4. (From [7], © 1987 IEEE. Reprinted with permission.)

CMOS, the supply current should consist only of the leakage current, which is typically very small.

A CMOS version of the above gate was simulated using SPICE. For a CMOS gate, the output voltage was found to be 3.08 V. As it is outside of the noise margin, it may or may not be recognized as logic 1. The fault has a more dramatic effect on the supply current. It changes from a normal value of 26 pA to 116 μA. This indicates that, in the presence of this fault, both n and p parts are on. Practically, all the current flows through the transistor NC. This contradiction between the switch-level and circuit-level analyses again suggests that the switch-level analysis cannot adequately represent the circuit behavior. As node 1 is connected to ground through a low resistance path and node 5 is connected to V_{dd} through a low resistance path, the voltage at node 5 is higher than that at node 1. There will be a current flow from node 5 to node 1 through NC, as a consequence of ARC phenomenon.

An n-MOS version of the gate was simulated under the same conditions. It also shows that, in the presence of the bridging fault, the n-network is on. This causes the output voltage to drop to about 0.93 V, which is recognized as logic 0 by the subsequent stages. The analysis also gives a current of 22 nA through the transistor NC, which contradicts the results of switch level analysis. These situations can be resolved by taking into account the ARC phenomenon.

Both examples given above show that if the source voltage of an n-MOS (or p-MOS) transistor becomes higher (or lower) than its drain voltage, the behavior of that transistor cannot be modeled at the switch level because of the ARC phenomenon. One can easily see that ARC can occur in an n-MOS gate only when the n-network is off. ARC can occur in the n-network (or p-network) of a CMOS gate only under the vectors for which it is normally off, and the complementary network is on. The bridging fault that leads to ARC can be easily detected by monitoring the power supply current. Some important observations regarding ARC are summarized as follows:

1. In an n-network or p-network, for any short between two internal nodes of a PCC, its behavior can be correctly modeled at the switch level.
2. In an n-network or p-network, for a short between two internal nodes of a PCC, there exists at least one vector for which the conductance state is different from that of the fault-free network.
3. In an n-network or p-network, for a short between two columns involving one nonlogical node within a column and one logical node (except ground), there is at least one vector under which the switch level analysis is not applicable.
4. In an n-network or p-network, for any short between two internal nodes of different PCCs, there exists at least one vector under which some transistors show ARC, and hence the switch level analysis is not applicable.

These observations identify the conditions under which switch-level analysis fails to predict the behavior of an n-network or p-network. A simple addition to the switch-level model to account for the ARC may be made by introducing a diode across the switch [7]. The diode represents the behavior of the substrate-drain junction. Under normal operation, when the drain voltage is higher than the source voltage, the diode is reverse-biased and the current through the switch is controlled by only the gate voltage. In case of a fault, if the source voltage becomes higher than the drain voltage, the diode becomes forward-biased and shows the effect of ARC. This modification can be introduced in the switch-level algorithms used for the simulation. Existing switch-level simulation algorithms may not support the representation of diodes without appropriate modifications.

3.8 SUMMARY

Bridging is a major failure mode in all electrical circuits. The effect of bridging faults is examined in detail for digital circuits. Static n-MOS and CMOS gates,

CMOS domino logic, CVSL gates, and clocked CMOS circuits are examined for possible bridging faults. The effect of bridging between logical nodes without feedback depends on the resistances involved and the noise margins. A resistance model is described for bridging faults without feedback. The model takes the value of bridging resistance and predicts the exact output voltage in the presence of a fault. For feedback bridging faults, oscillation may or may not occur depending on the delays involved. A necessary condition is obtained to predict the oscillations in the circuit in case of feedback bridging. It is shown that when a feedback bridging fault does not cause oscillations, it creates an anomalous output. Such faults are not testable by logic testing and significantly reduce the noise immunity. To detect such faults, current monitoring is recommended.

The accuracy of switch-level modeling for VLSI testing has been examined. SLA generally predicts the correct logic output, although it does not predict any degradation in noise margins. In some cases, it will predict a definite value, although the actual voltage stays outside the noise margin.

When the substrate is connected to the source, there are some cases in which the result from switch-level analysis does not match that of circuit-level analysis. For faults such as gate-to-source, gate-to-drain, and drain-to-source bridging, the switch-level model is adequate to obtain the necessary test vectors. For shorts between a logical input and nonlogical node of another column, the switch-level model does not correctly predict the behavior under all vectors. In these cases, an analysis at a level below the switch level is essential. Also, when a short occurs between two nodes of two different PCCs, the switch-level model may not be accurate enough to generate the necessary test vectors. Under some specific conditions, the ARC phenomenon must be taken into account.

Switch-level analysis can be used very effectively to characterize faults. However, in some specific cases, the circuit-level behavior must be taken into account for accuracy.

PROBLEMS

1. Explain why the gate level model for bridging fault as given in Figure 3.3 fails for input vector (1,1,1).

2. In a CMOS circuit, the on-resistance of n-MOS and p-MOS transistors are 3.0 kΩ and 7.5 kΩ, respectively. If the outputs of two inverters are bridged together with a 5 kΩ resistance, calculate the output voltages when one output is 1 and other is 0. What will be the output voltages if the bridge resistance is 50 kΩ? Assume the off-resistance of any transistor to be 10 MΩ.

3. Repeat Problem 2 for the circuit given in Figure 3.6 with inputs (i) $ABCD$ = 1000 and (ii) $ABCD$ = 1100.

4. Verify Equation (3.6) for oscillations in the feedback loop, which contains an odd number of inverters.

5. Give the complete test set for bridging faults shown in Figure 3.12. Also indicate which faults will not be deterministically detected by logic testing.

6. What is ARC? Simulate the circuit of Figure 3.17 using SPICE to show ARC in the presence of various bridging faults as shown in the figure.

REFERENCES

1. J.P. Shen, W. Maly, and F.J. Ferguson, "Inductive Fault Analysis of MOS Integrated Circuits," *IEEE Design and Test*, Dec. 1985, pp. 13–26.

2. Y.K. Malaiya, A.P. Jayasumana, and R. Rajsuman, "A Detailed Examination of Bridging Faults," *Proc. Int. Conf. Computer Design*, Oct. 1986, pp. 78–81.

3. Y.K. Malaiya and S.Y.H. Su, "A New Fault Model and Testing Technique for CMOS Devices," *Proc. Int. Test Conference*, 1982, pp. 25–34.

4. W. Maly and P. Nigh, "Built-in Current Testing—Feasibility Study," *Proc. Int. Conf. Comp. Aided Design*, 1988, pp. 340–343.

5. M. Shoji, *CMOS Digital Circuit Technology*, Prentice Hall, Englewood Cliffs, NJ, 1988.

6. R. Rajsuman, A.P. Jayasumana, and Y.K. Malaiya, "Testability Analysis for Bridging Faults in Dynamic CMOS," *Proc. Int. Symp. Electr. Dev. Cir. and Systems*, 1987, pp. 630–632.

7. R. Rajsuman, Y.K. Malaiaya, and A.P. Jayasumana, "On Accuracy of Switch Level Modeling for Bridging Faults in Complex Gates," *Proc. 23rd Design Auto. Conf.*, June 1987, pp. 244–250.

8. R. Rajsuman, "An Analysis of Feedback Bridging Faults in MOS VLSI," *Proc. IEEE VLSI Test Symp.*, 1991, pp. 53–58.

Chapter 4
Open Faults in Random Logic

4.1 INTRODUCTION

Fault modeling and testing of VLSI circuits have received considerable attention. Researchers have examined the fault mechanisms and the physical causes of failures in VLSI. Results reported are [1]:

1. Failures are randomly distributed in a VLSI system and all the blocks have equal probability of having a fault.
2. The nature of failures consists mainly of short and open circuits at the level of metallization and that of diffusion, but not between metallization and diffusion.

The same type of results have been reported in another study [2]. The major failure modes reported are short and open faults due to missing or excess metal (polysilicon). As reported, a major fraction of all faults are floating metal line faults. The presence of an open fault not only depends on the technology, but also on the mask topology.

This chapter describes the circuit behavior under different line open faults at transistor level. Testing methods to detect such faults are given. The problems in the testing are identified and possible solutions are mentioned.

4.2 MODELING OF OPEN FAULTS

In general, open faults are considered in the form of a complete break. In many cases, this assumption may not be correct. As an open fault is not an intentional feature, the break may not be well defined. A parallel combination of resistance and capacitance is used to model an open fault. A considerably high resistance (500 kΩ) is used to model the leakage current. The capacitance value is chosen depending upon the size of line break. A break in a typical 0.2 μm thick metal line can create a capacitance on the order of 10^{-15}F (line width about 20 μm). Most of the line open faults can be modeled by a capacitance ranging from 10^{-12}F to 10^{-15}F. Based upon the dielectric present in the break (air or silicon dioxide) and the line width, small deviation from this range is expected.

In n-MOS circuits a line open fault can always be modeled as a stuck-at fault. If an interconnection line from one gate to another is open, it can be modeled as an input s-a-0 fault for the second-level gate. If there is an internal open fault within a gate, it can be modeled as corresponding input's s-a-0 faults. An open fault in an n-MOS circuit can be detected by a test set developed to detect stuck-at faults. Such tests can be generated by standard techniques, i.e., D-algorithm or Boolean difference as discussed in Chapter 5.

Open faults in the CMOS circuits are considered nonclassical types of faults. The main reason is that, in the presence of an open fault, a CMOS combinational gate exhibits sequential behavior [3]. This sequential behavior cannot be modeled at the classical stuck-at level. This fact is illustrated by a simple two-input NAND gate as shown in Figure 4.1. In a fault-free gate, only one part—either the n-part or p-part—conducts for any set of input combinations. If an open fault exists in the series n-MOS (fault #1), the output node could never be pulled low. Thus, the fault is equivalent to the output stuck-at-1.

The interesting situation arises in the presence of an open fault in a parallel p-MOS network. Consider the fault #2 in Figure 4.1. In the presence of this fault, the output is prevented from being charged when the inputs are $mn = 10$. Because of the capacitive load, the output node retains previous value, or, in other words, the gate shows sequential behavior. This fault can be detected by a sequence of two vectors. The first pattern should initialize the output and the second vector sensitizes the fault. If an open fault occurs in the n-part (or p-part), the first pattern sets the output to logic 1 (logic 0). The second pattern then attempts to provide a

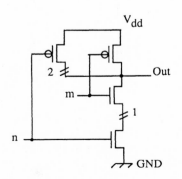

Figure 4.1 CMOS NAND gate to illustrate the sequential behavior in the presence of an open fault.

low resistance path between the output and ground (or power supply) through the faulty transistor.

The generation of two-pattern tests is a very complex process. This problem has been identified as Co-NP hard [4]. The requirement of large CPU time makes the test generation extremely costly. For a fault, a combinational block may not have any two pattern sequence. Furthermore, some problems have been identified to impose a limitation on the application of two-pattern tests. These problems with suggested solutions are discussed in the next section.

4.3 PROBLEMS IN TESTING OPEN FAULTS

In this section, we specify some problems that cause severe difficulties in the use of two-pattern sequences.

4.3.1 Test Invalidation by Timing Skews

During the transition from the first test pattern to the second, spurious logic values may invalidate the initialized value. In other words, stray circuit delays or timing skews may cause the failure of a two-pattern test. The situation is best illustrated by an example [7].

Example 1. Consider the following nonredundant combinational function:

$$F = AB + \overline{A} \cdot \overline{B} + CD + \overline{C} \cdot \overline{D} + AD$$

This function is implemented in two-level NAND-NAND form and AND-OR complex gate form, as shown in Figure 4.2(a). The Karnaugh map of the function is given in Figure 4.2b.

The PI AD covers only one EM, $A \overline{B} \overline{C} D$. Its nearest false vertices are $A \overline{B} C \overline{D}$ and $\overline{A} B \overline{C} D$, respectively. Both of the false vertices are two-Hamming distances away from the EM $A \overline{B} \overline{C} D$.

Let the p-MOS in the output NAND gate connected to line derived by PI AD be s-open. Then, to detect the fault a two-pattern sequence $\{\overline{A} B \overline{C} D, A \overline{B} \overline{C} D\}$ is used. However, due to circuit delays an intermediate state, i.e., $AB \overline{C} D$ or $\overline{A} B \overline{C} D$ might occur. Incidentally, these intermediate vectors are covered by some unaffected PI. Because of this, the circuit output will eventually become 1, irrespective of the presence of the fault. Thus, the presence of the fault may not be detected.

Note that among the *prime implicants* (PI), only the PI $'AD'$ does not cover any *essential minterm* (EM) that is adjacent to any false vertex of F. Hence, in either AND-OR complex gate or two-level NAND-NAND/NOR-NOR realization, the disappearance of AD due to some s-open fault cannot be tested in presence of circuit delays.

In the equivalent AND-OR complex gate, the equivalent s-open fault is marked (Figure 4.2(a)). To detect this fault, the initialization vector can be chosen from {(1010), (0110), (0101)}, while the second test vector is (1001). During the transition from the first vector to the second, an intermediate value (either 1011 or 1101) may charge the output node. To an observer, it will appear that the s-open fault is not present.

The above example illustrates a situation where all possible two-pattern tests may fail to detect the fault. In this situation, an IC tester will consider the circuit

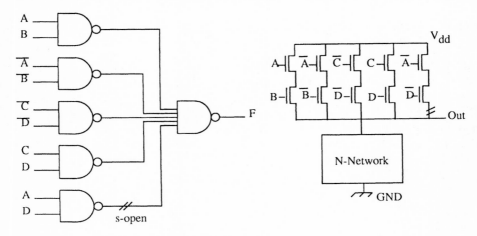

Figure 4.2(a) Combinational circuit used in Example 1.

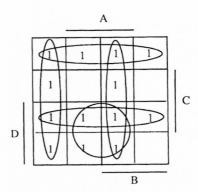

Figure 4.2(b) Karnaugh map for the function used in Example 1.

fault-free. However, as the fault remained undetected due to random delays, the circuit will not work as part of a printed circuit board. This problem can be avoided if the Hamming distance between the initialization vector and sensitization vector is kept at unity [5–6].

4.3.2 Test Invalidation by Charge Distribution

Another problem in the detection of s-open fault is the redistribution of charge among the parasitic capacitances associated with internal nodes. In the presence of an open fault, during second test vector, the output node of a CMOS gate shows the high impedance state. The charge retained in the load capacitance during this high impedance state determines whether a fault is present. During this high impedance state, the charge stored in the load capacitance is shared by the parasitic capacitances of internal nodes. If the gate under test is a large complex gate, which implies that there are several internal nodes, charge redistribution may result in a significant voltage drop at the output. If the charge remaining in the load capacitance is small, the voltage level will fall outside the noise margin. In this situation, the next stage may fail to recognize the correct logic value. Consequently, to an observer, the gate under test will appear to be fault free [5].

Note that this problem is dependent on the function being realized in actual implementation, physical topology as well as the input pattern. In general, for primitive gates (NAND, NOR, and NOT), the probability of test invalidation because of charge distribution is negligible. However, the problem cannot be neglected in complex gates.

4.3.3 Test Invalidation Due to Glitches

Another problem in detection of s-open faults is the presence of *glitches* (hazards) in the circuit. There are many reason of formation of glitches. They may also occur due to delays in the circuit along various paths. Other reasons might be the switching of some transistors, external electromagnetic interference, and ionization radiation. The formation of glitches may be avoided in specific cases. In general, however, to identify every possible glitch and redesign the circuit to avoid all of them would not be feasible. The failure of a test sequence due to glitches can be explained in a manner similar to the way in which charge distribution causes the test to fail. Testability of a circuit is affected drastically in the presence of glitches [8–10]. The situation is best illustrated with an example.

Example 2. To illustrate the failure of test sequences in the presence of glitches, consider the CMOS complex gate shown in Figure 4.3. If I is an input vector, the gate implements the function $\overline{AB + CD}$, which is $I \rightarrow (1,0)$.

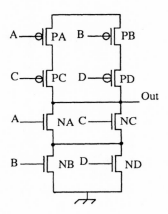

Figure 4.3 CMOS example gate to show the failure of two-pattern test sequences in the presence of glitches.

Consider the fault, transistor NA stuck-open. To detect this fault, a test T is used, where $T = (T_1, T_2)$. Here T_1 is the initialization vector and T_2 is the actual test which sensitizes the fault. The initialization pattern can be chosen from the set:

$$[ABCD] = [0101, 0001, 0100, 1010, 0010, 1000, 0000]$$

The second pattern of the sequence can be chosen from the set:

$$[ABCD] = [1100, 1001, 1101]$$

To avoid the potential invalidation of the test sequence due to timing delays, a sequence which has unity Hamming distance between its patterns can be chosen from the following set:

$$[ABCD, ABCD]$$

$$= [(0100, 1100), (1000, 1100), (0001, 1001), (1000, 1001), (0101, 1101)]$$

The first pattern sets the output node to logic 1 and the second pattern creates a high impedance state in the presence of the fault, i.e.,

$$T \rightarrow (1, Z)$$

A positive glitch (0-to-1 G) at node C, caused by the prior logic, may turn on transistor NC momentarily during the second pattern. This will create a low resistance path from the output to ground. If the glitch is sufficiently wide, the charge at the output node may conduct to ground. This effect will appear as

$$T \rightarrow (1, GZ) \rightarrow (1,0)$$

Consequently, the test sequence fails to detect the fault.

This example shows that any possible test sequence may be invalidated and the fault could remain undetected. Similarly, one can show that the negative glitches (1-to-0 G) may cause the failure of all possible test sequences for stuck-open fault in the p-part. The reason for the failure of a sequence is the presence of glitches that interfere with the high impedance state. Hence, the failure of a sequence is possible, irrespective of the method used for initialization. Such failures are possible not only for two-pattern sequences, but also for multiple-pattern sequences. These observations are summarized as follows:

1. In a CMOS gate, except for a NOT gate, in at least one part (either the n-part or p-part), all FET stuck-open faults may remain undetected in the presence of circuit glitches.
2. One cannot have a CMOS gate, other than an inverter, in which two-pattern or multipattern tests for stuck-open faults are guaranteed to remain valid in the presence of glitches.

Note that if a test sequence remains valid in the presence of glitches for a FET stuck-open fault, any possible sequence for the dual transistor may be invalidated because of glitches. The reason can be explained with a CMOS structure. A sequence remains valid for a stuck-open fault in one part if and only if there is only one possible conduction path from the output to the ground or power supply. However, in the dual part, there will be more than one conduction path from the output to the power supply or ground. Glitches may spuriously switch an undesirable path in the dual part and cause the failure of test sequence.

4.4 METHODS TO TEST STUCK-OPEN FAULTS

In Section 4.3, we discussed problems in the testing of open faults. To overcome these problems, two solutions have been developed. The first solution is to use specially designed test sequences, and the second solution is to design (or redesign) the circuit in a special manner. In this section, we will describe these solutions.

4.4.1 Robust Test Sequences

To overcome the problem of potential test invalidation of two-pattern or multipattern tests, special test sequences are used. In general, these special test sequences are known as *robust test sequences* (RTS). The criteria of robustness for a test pair are given as follows:

1. The Hamming distance between the initialization vector and the sensitization vector should be unity. Thus, the possible intermediate state is avoided.
2. During the application of test pair, the sensitized path is free from any circuit glitches.

These requirements impose difficult limitations on a test sequence to be robust. The first criterion implies that in every sum-of-product expression the primary implicant should have at least one essential true vertex that is adjacent to a false vertex. In general, a circuit does not satisfy this criterion. By introducing redundancy in the circuit, this requirement can be satisfied. However, a redundant circuit cannot be tested for 100% fault coverage.

The second criterion is even harder to satisfy. It implies that if (T_1, T_2) detects a s-open fault in the p-part (or n-part), the output of the gate under test should never become 1 (or 0) due to a glitch. As mentioned above, glitches may occur due to various reasons. To identify all glitches and redesign a circuit to avoid them would be impossible. In [11], a method is suggested to check the robustness of a test sequence. The basic idea is to monitor all transitions at all lines in the circuit during (T_1, T_2), while T_1 and T_2 are unit Hamming distance apart. If any signal has more than one transition, a glitch is detected and the test is called *nonrobust*. To detect a glitch on a line during test application, T_1 and T_2 are merged into T. If the jth literal in T_1 and T_2 is represented as $T_1(j)$ and in $T_2(j)$, respectively, the merging rules are [11]:

1. If $T_1(j) = T_2(j) = 0$; then $T(j) = 0$ or $0 - G$.
2. If $T_1(j) = T_2(j) = 1$; then $T(j) = 1$ or $1 - G$.
3. If $T_1(j) = 0$, and $T_2(j) = 1$; then $T(j) = R$, represents a 0-to-1 transition.
4. If $T_1(j) = 1$, and $T_2(j) = 0$; then $T(j) = F$, represents a 1-to-0 transition.

If a merge operation is performed on T_1 and T_2 for all literals and, for all j, $T(j)$ is either 0 or 1, the test sequence is robust. If, for any j, $T(j)$ has a glitch (either $0 - G$ or $1 - G$), R or F, the test sequence is not robust.

Because of the above-mentioned limitations, to calculate robust sequences for all s-open faults in a circuit is extremely difficult. In [11], an algorithm is given to calculate robust sequences. However, for many circuits, no robust sequences are found for many s-open faults.

4.4.2 Testable Designs

As discussed in Section 4.4.1, the generation of robust test sequences is an extremely complex process. The requirement of large CPU time makes the test generation excessively costly. Furthermore, in a large number of cases, a combinational block does not have any robust sequence. To overcome this problem, testable designs are suggested. These designs use extra transistors controlled by external signals to obtain a multipattern test sequence.

4.4.2.1 Designs Using Multipattern Sequences

In Figure 4.4, basic concepts of testable gates are shown. In all the circuits of Figure 4.4, a two-pattern test sequence can be obtained by appropriately con-

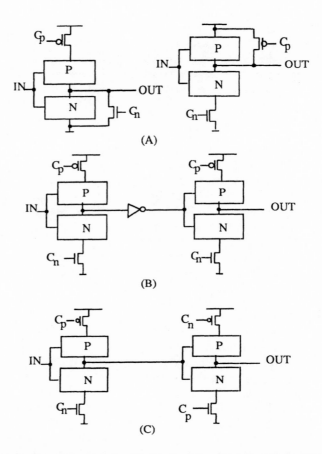

Figure 4.4 Testable CMOS designs: (a) after [5], (b) after [6], and (c) after [7]. (© 1991 IEEE. Reprinted with permission.)

trolling the signals Cp and Cn. In Figure 4.4(a), two extra MOSFETs are used, controlled by external signals Cp and Cn ($Cp = Cn$) [5]. One n-MOS (p-MOS) is used in series with the n-part (p-part), while other p-MOS (n-MOS) is used in parallel with the p-part (n-part). In structure (i) of Figure 4.4(a), a two-pattern robust sequence can be obtained for any p-MOS s-open fault, while in structure (ii) a test sequence can be obtained for n-MOS s-open fault. For example, to test a p-MOS s-open fault in structure (i), a proper initializing vector $T1'$ and test vector $T2'$ are $T1' = T2$, control signal $= 1$, and $T2' = T2$, control signal $= 0$, where $T2$ is the sensitization vector for original gate.

In Figure 4.4(b), two MOSFETs are used to make the gate testable. One p-MOS and an n-MOS are used in series with the p-part and the n-part, respectively [6]. An additional inverter is used at the output of every gate to propagate the fault effect through the circuit. This inverter is eliminated in Figure 4.4(c) by adjusting the control signal in successive stages [7]. In this design, the "all 1s" or "all 0s" vector is used as initialization vector, depending upon whether the fault under test is in the p-part or n-part, respectively. The second vector is either a 1_v with $Cp\ Cn = 01$ (to test the p-part) or a 0_v with $Cp\ Cn = 01$ (to test the n-part).

The designs given in Figure 4.4 are effective to a certain extent, but do not overcome the problem due to glitches [8–10]. In the presence of glitches, a test may fail to detect an open fault even in the testable designs of Figure 4.4. The main reason of test invalidation is the presence of high impedance state at the output during last vector. The charge at the output during the high impedance state determines whether a fault is present. As long as the sequential behavior of the CMOS gate is used for stuck-open fault detection, a test may invalidate. To avoid this problem, a faulty circuit should be kept combinational during testing. This implies that the high impedance state should not be allowed during testing. In other words, the testing procedure used for n-MOS technology, which uses single test vectors, should be used for the testing of CMOS gates.

4.4.2.2 Design Using Single Test Vector

Testing a CMOS gate using single test vector requires the output of the gate to be connected to the power supply (ground) during testing of the n-part (p-part). The resistances of these connections should be considerably higher than the on-resistances of the n-part or p-part. In fact, resistance ratios should be made similar to the resistance ratio of the load (depletion transistors) and on-resistances of the network in n-MOS gates.

In CMOS, both n-part and p-part are complementary. If a vector attempts to switch on the p-part, it switches off the n-part and *vice versa*. Hence, we need to add a p-type transistor Tp parallel to the p-part, and an n-type transistor Tn

parallel to the *n*-part, as shown in Figure 4.5. These extra FETs require two additional control signals, *Cp* and *Cn*, which keep them off during normal operation, but on during testing. During normal operation $Cp = 1$ and $Cn = 0$. During testing of the *n*-part $Cp = Cn = 0$ and during testing of the *p*-part $Cp = Cn = 1$. The gate is essentially transformed to a pseudo-n-MOS gate (pseudo-p-MOS gate), when testing for stuck-open faults in the *n*-part (*p*-part) as shown in Figure 4.6. The standard rules to design pseudo-n-MOS (pseudo-p-MOS) gates can be used to determine the sizes of *Tp* and *Tn*, respectively [10]. By equating current through the *n*-part and *p*-part, the β_n/β_p of *Tp* and *Tn* can be estimated as 1/6. In practice, some deviation from this ratio of 1/6 is acceptable. This ratio should be chosen to obtain minimum area while considering the transfer characteristics of the gate without significantly degrading the noise margins. Based upon SPICE simulations, a possible range is given as 1/10 to 1/2.

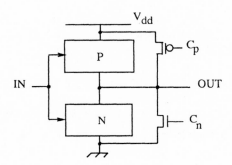

Figure 4.5 Design of single pattern testable CMOS gate. (From [7], © 1989 IEEE. Reprinted with permission.)

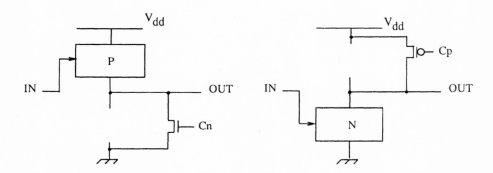

Figure 4.6 Transformation of CMOS gate into pseudo-p-MOS or a pseudo-n-MOS gate during testing. Under 0_r if the *p*-part is off, and under 1_r if *n*-part is off.

Using the above guideline for $\beta_n/\beta_p = 1/6$, the size of Tp and Tn can be calculated and given in Table 4.1 for NAND and NOR gates. Note that a NOT gate does not need an extra transistor, a NAND gate needs only Tn, and a NOR gate needs only Tp.

The design given in Figure 4.5 is not practical due to the requirement of two transistors per gate plus the routing of control signals. The extra hardware results in a significant penalty in area. Secondly, because of two extra transistors at the output, the output capacitance of the gate increases. This degrades the high-to-low as well as low-to-high switching delay. In high performance circuits, this excessive delay is highly undesirable.

The functionality of the design given in Figure 4.5 can be achieved by a single transistor instead of using two transistors. The switching of this transistor is controlled by an external signal Sc and the value passed is provided by the signal St. The concept is shown in Figure 4.7 [10]. If an n-MOS transistor is used, during normal operation, Sc is set to 0. However, if a p-MOS transistor is used, during normal operation, Sc is set to 1. The advantage of using n-MOS transistor is that the size of transistor is small compared to the p-MOS while offering the same resistance.

By appropriately controlling the signal Sc and St, the gate can be transformed into pseudo-p-MOS or pseudo-n-MOS gate. Hence, single test vector is used to detect a s-open fault in the functional part.

Some useful observations for this design can be summarized as follows:

1. By augmenting a CMOS combinational gate as shown in Figure 4.5, any single stuck-open fault in the functional part can be detected by a single test vector. These tests are not invalidated by timing skews and delays, glitches, or charge-sharing among the internal nodes.
2. An augmented CMOS gate, as shown in Figure 4.5, can be tested for all single stuck-open faults in the functional part by a sequence of maximal length

Table 4.1
Suggested sizes of extra transistors for primitive gates; NA represents that the corresponding transistor is not required. The calculations are for 2 μm technology, minimum p-MOS size $L = 2\mu$m, $W = 12\mu$m, and minimum n-MOS size $L = 2\mu$m, $W = 6\mu$m.

Gate	Tp size	Tn size
Inverter	NA	NA
2 input NAND	NA	$L = 4u, W = 2u$
3 input NAND	NA	$L = 6u, W = 2u$
2 input NOR	$L = 2u, W = 4u$	NA
3 input NOR	$L = 2u, W = 6u$	NA

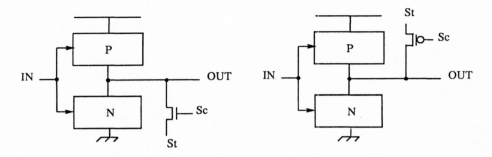

Figure 4.7 Efficient implementation of single pattern testable CMOS gates. (From [10], © 1991 IEEE. Reprinted with permission.)

2n, where n is the number of transistors in the unaugmented n-part or p-part.
3. In the augmented CMOS gate of Figure 4.5, a test set which detects all single stuck-open faults in the functional transistors will also detect all multiple stuck-open faults in the functional part.

4.4.2.3 Comparison and Advantages of Testable Designs

The most important aspect to choose a testable design is the amount of extra hardware. The last design is better than any other design in this respect. Also, due to the use of a single transistor at the output, the increase in output capacitance is negligible in the latter design. While all other designs have significant penalty in switching speed due to the use of two transistors, the last design offers almost no penalty in switching speed. The testable designs are compared in Table 4.2.

The major advantage of the CMOS testable design given above is that any stuck-open fault can be detected by a single test vector. This reduces the testing

Table 4.2
Comparison of testable designs to make CMOS circuits s-open fault testable.

Schemes	# of Vectors	Extra Hardware	Problems by Glitches
Ref. [5]	2	2 FETs + 2 controls	yes
Ref. [6]	2	2 FETs + 2 controls + 1 inverter	yes
Ref. [7]	2	2 FETs + 2 controls	yes
Ref. [9]	1	2 FETs + 2 controls	no
Ref. [10]	1	1 FET + 2 controls	no

time drastically. First, it reduces the test application time by 50% because it requires a single test vector instead of a sequence of two vectors. Furthermore, it eliminates the complexity in test generation. Complexity in generating two-pattern or multi-pattern sequences is a major cost factor in testing CMOS stuck-open faults. The complexity and cost associated with the generation of robust test sequences is even higher.

As only a single pattern is required to test a given fault, the tests for the augmented gates can be generated by any standard procedures. All the classical algorithms such as the D-algorithm, and *automatic test pattern generating programs* (ATPGs) for n-MOS can generate the tests for such augmented gates. The scheme detects the stuck-open faults deterministically and ensures the detection irrespective of the problems discussed in Section 4.3. Also, this design offers significant advantage for random or pseudorandom testing procedures. Such testing is very inefficient for the detection of stuck-open faults because the probability of testing a fault depends on two successive vectors.

Some designs, such as those given in Figure 4.4, are not suitable for multilevel circuits. This is mainly due to the problems associated with the propagation of the fault effect to the circuit output. The design given in Figure 4.7 is free from this drawback. As a single test vector is used and no part is intentionally kept off, the effect of a fault is propagated without any problem. In fact, testing of CMOS gates designed by this procedure is comparable with the testing of n-MOS gates. To illustrate, consider the circuit shown in Figure 4.8 [5]. To detect stuck-open fault in the p-FET driven by the open line requires a three-pattern sequence: $[ABCD]$ = [1001, 1011, 1010]. However, using the design of Figure 4.7, the same fault can be detected by a single vector $[Sc\ St\ ABCD]$ = [101010].

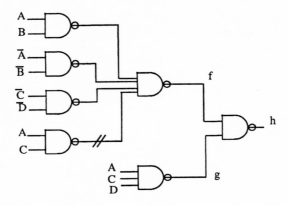

Figure 4.8 CMOS combinational circuit for which no two-pattern sequence exists having vectors that are unit distance apart. (From [5], © 1986 IEEE. Reprinted with permission.)

A disadvantage of the design in Figure 4.7 is the slightly higher power dissipation during testing. However, this is not a significant disadvantage, as it occurs only during testing. In normal operation, the extra transistor is kept off.

4.5 TESTABILITY OF DYNAMIC CIRCUITS

In the above sections, we have examined testability of static CMOS gates. In this section, we discuss the testability of dynamic gates for open faults.

Dynamic circuits have always interested circuit designers due to their high switching speed, synchronization capability, and requirement of fewer transistors. In CMOS technology, two logic families, domino CMOS and clocked CMOS, are very popular. With the growing need for high performance circuits, a large number of chips have been designed by using domino and clocked CMOS logic. This section examines testability of these circuits. Other dynamic circuits, i.e., zipper CMOS, DCVS, and NORA can be derived from domino CMOS, and hence these are not discussed.

In clocked CMOS circuits, two extra transistors are used with static CMOS structure. One clocked transistor (n-MOS) is used in series with the *n*-part and the other clocked transistor (p-MOS) is used in series with the *p*-part. If the clock is considered as an external control signal, the structure appears as given in Figure 4.4(b). The testability of this design is discussed in Section 4.4, and hence it can be neglected here.

In domino CMOS circuits, the function is implemented by the *n*-type driver part, which also associated with a clocked n-MOS transistor. The load part consists of a single clocked p-MOS transistor. Each such structure is followed by a static CMOS inverter to propagate the signal from one stage to another (Figure 4.9). When the clock is 0, the input node of static CMOS inverter is precharged to 1. This precharged node is conditionally discharged during evaluation phase (clock is 1). Because of the static inverter, the gate output and hence the inputs of the successive gates remain low during the precharge phase.

The testability of domino logic has been examined in [12–13], which report that an open fault in the clocked transistors or an open fault in the static inverter can be modeled as stuck-at fault [13]. An open fault in the *n*-type driver part creates a high impedance state at the inverter input during the evaluation phase. Hence, to test such a fault, a two-pattern sequence is required. In domino logic, utilizing the structural properties, a two-pattern sequence can be easily obtained. Because all the inputs remain low during the precharge phase, it is used for initialization purposes. The second test vector, which is calculated for an equivalent n-MOS circuit, is applied during the evaluation phase. Hence, the complete sequence is applied within a single clock period. It should be noted that the gate inherently has a precharge phase. Therefore, the test generation complexity does not increase

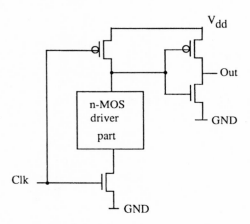

Figure 4.9 A generalized CMOS domino gate.

due to the requirements of two-pattern tests. Also, all inputs remain 0 during the precharge phase, and only unidirectional transitions are allowed during evaluation phase. Hence, a domino circuit does not suffer the problems discussed in Section 4.3. The two-pattern sequence, which is applied within a clock period, cannot be invalidated by timing skews or circuit glitches in domino CMOS circuits.

4.6 SUMMARY

Stuck-open faults present severe testing problems in CMOS circuits. Two-pattern or multipattern sequences were observed to fail to detect the fault. Robust test sequences, and even the augmented gates which utilize high impedance state, are not suitable for deterministically detecting the fault. The domino CMOS circuits are free of such problems.

A testable design is given to overcome testing problems. The design uses a static load at the output and allows testing it like a pseudo-p-MOS or n-MOS gate. The design requires minimal extra hardware. The test phase for this design is simple and uses a single test vector to detect a fault. In this design, tests are not invalidated due to timing skews and delays, glitches, or charge-sharing among the internal nodes.

PROBLEMS

1. Identify all possible s-open faults in Figure 4.1. Obtain the test vectors to detect these faults.

2. Give the test vectors to detect faults #1 and #2 as shown in Figure 4.1. Model these faults with a parallel combination of resistance and capacitance, and simulate the circuit with SPICE.

3. Give all possible robust test sequences for a CMOS complex gate that implements $\overline{(AB + C) \cdot (DE + F)}$.

4. Make a list of test vectors to cover all single transistor s-open faults in Figure 4.7. Modify the circuit with efficient augmentation method given in Section 4.4.3 and update the list of test vectors.

5. Repeat problem 3 for Figure 4.8.

6. Why cannot a two-pattern sequence within a clock period designed to detect an open fault be invalidated in domino CMOS circuits?

REFERENCES

1. J. Galiay, Y. Crouzet, and M. Vergniault, "Physical versus Logical Fault Models in MOS LSI Circuits and Impact on Their Testability," *IEEE Trans. Comp.*, **c-24**, June 1980, pp. 527–531.

2. W. Maly, "Modeling of Lithography Related Yield Losses for CAD of VLSI Circuits," *IEEE Trans. CAD*, **CAD-4**, July 1985, pp. 166–177.

3. R.L. Wadsack, "Fault Modeling and Logic Simulation of CMOS and NMOS Integrated Circuits," *Bell Sys. Tech. J.*, **57**, May–June 1978, pp. 1449–1474; "Fault Coverage in Digital Integrated Circuits," pp. 1475–1488.

4. S.K. Chakravarty, "On the Complexity of Computing Tests for CMOS Gates," *IEEE Trans. CAD*, **8**(9), Sept. 1989, pp. 973–980.

5. S.M. Reddy and M.K. Reddy, "Testable Design for CMOS Logic Circuits," *IEEE Trans. Comp.*, **c-35**, Aug. 1986, pp. 742–754.

6. D.L. Liu and E.J. McCluskey, "CMOS Scan Path IC Design for Stuck Open Fault Testability," *IEEE J. Solid State Cir.*, **22**(5), Oct. 1987, pp. 880–885.

7. B. Gupta, Y.K. Malaiya, Y. Min, and R. Rajsuman, "On Designing Robust Testable CMOS Combinational Circuits," *IEEE Proc.*, part E, **136**(4), July 1989, pp. 329–338.

8. R. Rajsuman, Y.K. Malaiya, and A.P. Jayasumana, "CMOS Stuck-Open Fault Testability," *IEEE J. Solid State Circuits*, **24**(1), Feb. 1989, pp. 193–194.

9. R. Rajsuman, A.P. Jayasumana, and Y.K. Malaiya, "CMOS Open Fault Detection in Presence of Glitches and Timing Skews," *IEEE J. Solid State Circuits*, **24**(4), Aug. 1989, pp. 1055–1061.

10. A.P. Jayasumana, Y.K. Malaiya, and R. Rajsuman, "Design of CMOS Circuits for Stuck Open Fault Testability," *IEEE. J. Solid State Circuits*, **26**(1), Jan. 1991, pp. 58–61.

11. S. Chakravarty, "A Characterization of Robust Test-Pairs for Stuck-Open Faults," *J. Electronic Testing*, **2**(1), Feb. 1991, pp. 275–286.

12. V.G. Oklobdzija and P.G. Kovijanic, "On Testability of CMOS Domino Logic," *Proc. Int. Sympt. Fault Tolerant Computing*, 1984, pp. 50–55.

13. R. Rajsuman, A.P. Jayasumana, and Y.K. Malaiya, "Testability Analysis for Bridging Faults in Dynamic CMOS," *Proc. Int. Symp. Electr. Dev. Cir. and Sys.*, 1987, pp. 630–632.

Chapter 5
Test Generation and Fault Simulation

5.1 INTRODUCTION

In small geometry VLSI circuits, fault detection is a very complicated problem. Due to micron (or submicron) level geometries, an arbitrary node cannot be physically probed in a VLSI circuit. The only accessible points are the input and output pins. By applying some binary value at the inputs and examining the response at the outputs, one should deduce whether an embedded gate (or line) has a fault. Another problem is the length of a test set. A VLSI circuit cannot be tested under all possible input combinations (called *exhaustive testing*). An n-input circuit can have 2^n possible input combinations. In VLSI circuits, n can easily be in the range of 25 to 50. A tester running at 10 MHz frequency will require about 3.6 years to generate 2^{50} test vectors. Hence, in test generation, an important criterion is to minimize the length of a test set. The total test time includes the time of test generation, test application, and that required to evaluate the response.

Several algorithms have been developed to generate test vectors for a combinational logic circuit. The basic objective is to obtain a set of input combinations for which the fault-free and faulty responses of the circuit are different. Algorithms based on a gate-level description of the circuit are very popular. These algorithms do not cover physical defects and are limited to stuck-at faults. Because of the ease in test generation and reasonable fault coverage, these algorithms are widely used. In this chapter, we will describe the most popular test generation algorithms for combinational circuits and provide a comparison. The methods for fault simulation and how to obtain an estimate for fault coverage are also discussed.

5.2 TEST GENERATION AT GATE LEVEL

The objective of test generation is to obtain a set of input vectors which will produce an opposite logic value at the output in the presence of a fault as compared to the fault-free output. The aim is just to detect the presence of a fault, fault location being a different and much more complex problem.

All the methods developed to generate test vectors work fairly well under the following assumptions:

1. The *circuit under test* (CUT) has no redundancy.
2. The circuit has only one stuck-at fault at a time.

The second assumption (known as the *single stuck-at fault assumption*) is unrealistic because faults are found in clusters. However, the test set developed to cover all single stuck-at faults also provides good coverage for multiple faults. Hence, to minimize the testing cost, the single fault assumption has been widely used. The first assumption is necessary because a redundant fault does not cause an error, and hence is not detected. We have mentioned that this text is limited to static faults and discussion on intermittent faults is beyond the scope of this book.

5.2.1 Boolean Difference Method

The idea behind the *Boolean difference method* is that a fault can be detected if the faulty response of the circuit is different from the fault-free response. Assume that a circuit implements a Boolean expression $F(X)$ with n literals, i.e., $F(x_1, x_2, \ldots, x_i, \ldots, x_n)$. If the response under fault, line x_i stuck-at is $F(x_1, \ldots, \bar{x}_i, \ldots, x_n)$, the fault can be detected if

$$F(x_1, \ldots, x_i, \ldots, x_n) \text{ XOR } F(x_1, \ldots, \bar{x}_i, \ldots, x_n) = 1$$

This XOR operation is known as the Boolean difference of $F(X)$ with respect to input x_i [1], and is represented as $dF(X)/dx_i$. The Boolean difference $dF(X)/dx_i = 0$ represents that the faulty and fault-free responses are the same, and hence, a fault cannot be detected.

Some properties of the Boolean difference are given as [1]:

$$\frac{dF(X)}{dx_i} = \frac{\overline{dF(X)}}{dx_i} \tag{5.1a}$$

$$\frac{dF(X)}{dx_i} = \frac{dF(X)}{d\bar{x}_i} \tag{5.1b}$$

$$\frac{dF(X)}{dx_i} = 0; \text{ if } F(X) \text{ is independent of } x_i \tag{5.2a}$$

$$\frac{dF(X)}{dx_i} = 1; \text{ if } F(X) \text{ depends only on } x_i \tag{5.2b}$$

$$\frac{dF(X)G(X)}{dx_i} = F(X) \cdot \frac{dG(X)}{dx_i} \text{ XOR } G(X)$$
$$\cdot \frac{dF(X)}{dx_i} \text{ XOR } \frac{dF(X)}{dx_i} \frac{dG(X)}{dx_i} \tag{5.3}$$

$$\frac{dF(X) + G(X)}{dx_i} = \overline{F(X)} \cdot \frac{dG(X)}{dx_i} \text{ XOR } \overline{G(X)}$$
$$\cdot \frac{dF(X)}{dx_i} \text{ XOR } \frac{dF(X)}{dx_i} \cdot \frac{dG(X)}{dx_i} \tag{5.4}$$

After calculation of Boolean difference, the test vectors can be obtained as follows:

Test vector for x_i s-a-0 fault is: $x_i \cdot \dfrac{dF(X)}{dx_i}$ \hfill (5.5a)

Test vector for x_i s-a-1 fault is: $\overline{x}_i \cdot \dfrac{dF(X)}{dx_i}$ \hfill (5.5b)

The procedure is best illustrated with an example.

Example 1. Consider the circuit shown in Figure 5.1. For this circuit, output Z is given as $Z = ABC + B\overline{D} + ABD$. The Boolean difference with respect to line A is given as

$$\frac{dZ}{dA} = \frac{d(ABC + B\overline{D} + ABD)}{dA}$$

$$= \overline{(ABC + ABD)} \cdot \frac{d(B\overline{D})}{dA} \text{ XOR } \overline{(B\overline{D})}$$

$$\cdot \frac{d(ABC + ABD)}{dA} \text{ XOR } \frac{d(B\overline{D})}{dA} \cdot \frac{d(ABC + ABD)}{dA}$$

$$= [\overline{B} + D] \cdot \frac{d(ABC + ABD)}{dA}$$

$$= [\overline{B} + D] \cdot \left[A \cdot \frac{d(BC + BD)}{dA} \text{ XOR } (BC + BD) \right.$$

$$\left. \cdot \frac{dA}{dA} \text{ XOR } \frac{d(BC + BD)}{dA} \cdot \frac{dA}{dA} \right]$$

$$= [\overline{B} + D] \cdot [BC + BD]$$

or

$$\frac{dZ}{dA} = BCD + BD = BD$$

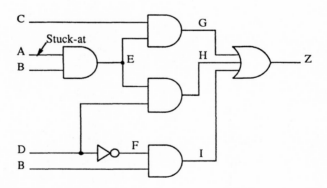

Figure 5.1 Combinational circuit for Example 1.

This means, a fault on line A can be observed if $B = 1$ and $D = 1$. The test vector to detect line A s-a-1 fault is $ABCD = 01x1$, and for fault line A s-a-0 is $ABCD = 11x1$, where x represents a "don't-care" value.

The above example illustrates the test generation for an input line. If the objective is to generate a test vector for an intermediate line, the output expression should be written in terms of that intermediate line. Then, the Boolean difference of the output can be obtained with respect to that line.

Example 2. To get a test vector for line E stuck-at fault in Figure 5.1, the output Z can be written as $Z = EC + ED + B\overline{D}$. The Boolean difference with respect to line E is given as

$$\frac{dZ}{dE} = \frac{d[(EC + ED) + B\overline{D}]}{dE}$$

$$= [\overline{B} + D] \cdot \frac{dE \cdot (C + D)}{dE}$$

$$= [\overline{B} + D] \cdot [C + D] = \overline{B}C + D$$

Thus, the test for E s-a-0 is: $E \cdot [\overline{B}C + D] = AB \cdot [\overline{B}C + D] = ABD$, or $11x1$. The test for E s-a-1 is: $\overline{E} \cdot [\overline{B}C + D] = (\overline{A} + \overline{B}) \cdot (\overline{B}C + D)$. This gives multiple test vectors $\overline{A}\overline{B}C + \overline{A}D + \overline{B}C + \overline{B}D$. Any vector from set $[001x, 0xx1, x01x, x0x1]$ can be used to detect line E s-a-1 fault.

Boolean difference method guarantees to find a test vector if a test exists. In general, $dF(X)/dx_i = 0$ detects a redundancy in the circuit. One advantage of this method is that it finds all possible test vectors for a fault. The algorithm is deterministic, but it requires excessively large amounts of time to compute these test vectors. The method is not practical, even for moderately sized circuits having more than 1000 gates.

5.2.2 Path Sensitization and D-Algorithm

The simplest path sensitization procedure is the single-path sensitization. This procedure selects a single path through which a fault effect can propagate. This path is called *sensitized* if a set of input values can propagate the fault effect to an observable or primary output. A major restriction in this procedure is that only one path is sensitized at a time from the faulty line to a primary output.

Example 3. Consider the circuit given in Figure 5.2. To test line C stuck-at-1 fault, line D must be set to 1 and line C should be 0. Thus, in the presence of a fault, line F becomes 1; otherwise, $F = 0$. This fault effect can propagate either through the path FGZ or the path FHZ. Suppose we select the path FGZ. The fault effect can propagate to line G by setting line E to 0. Finally, the fault effect propagates to the output Z by setting line H to 0. This process of propagating the fault effect is called *forward trace*. To obtain a test vector, the values at lines E and H must be obtained (called *line justification*) by assigning appropriate values at the inputs. In this case, line E can be set to 0 by setting line A to 1, and line H

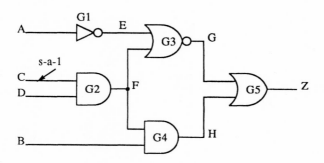

Figure 5.2 Combinational circuit to illustrate the path sensitization procedure.

can be set to 0 by setting line *B* to 0. Thus, a test vector for line *C* s-a-1 fault is
ABCD = 1001.

This example illustrates an important concept of propagating a fault effect.
If the fault effect needs to be propagated from an input to the output of a NOR/
OR gate, all other inputs to that gate should be set to 0. If a fault effect needs to
be propagated to the output of a NAND/AND gate, all other inputs to that gate
should be set to 1. These rules are summarized in Table 5.1.

Table 5.1
Rules to propagate the fault effect to the output of a gate.

Type of Gate	Values on Other Inputs
AND	all 1s
NAND	all 1s
OR	all 0s
NOR	all 0s

The application of the single-path sensitization procedure is limited due to
the restriction that a fault effect is propagated through only one path. In many
cases, sensitizing more than one path becomes necessary to drive a test vector.
This restriction is overcome in the D-algorithm, where the fault effect propagates
through all possible paths [2]. The basic algorithm is given as follows:

Step 1: Assume a value *D* or $\overline{D}$ at the line for which a test vector is needed.
This value represents the fault-free state of the line. The value *D* (or $\overline{D}$) can be

chosen as either 0 or 1, but should be opposite to the stuck-at value. Also, the value of D (or $\overline{D}$) should be consistent during the whole calculation. For example, if D is chosen to represent a 1, $\overline{D}$ is 0 throughout the circuit and *vice versa*.

Step 2: Make a list of all possible paths which can be sensitized from the fault-site to a primary output.

Step 3: Select one path at random, sensitize it, and delete it from the list. This means that the value D propagates through the circuit until it reaches a primary output. The propagation rules are the same as given in Table 5.1. This step is equivalent to *forward trace*, and is called "D-drive."

Step 4: Check for consistency of values assumed on various lines during D-drive. This step is equivalent to line justification. If an inconsistency is observed, repeat step 3 and propagate the fault effect through a different path. This is called *backtrack* and this process continues until there is a name in the path list.

Step 5: A successful consistency check provides a test vector. If, for all possible sensitizing paths, the consistency check fails, the fault cannot be detected. In general, this represents a redundancy in the circuit.

Example 4. This procedure can be illustrated by Figure 5.2. Consider the fault, line C s-a-1. The various steps are given as follows.

D-drive Operation:

Gate	A	B	C	D	E	F	G	H	Z
G2			$\overline{D}$	1		$\overline{D}$			
G3			$\overline{D}$	1	0	$\overline{D}$	D		
G5			$\overline{D}$	1	0	$\overline{D}$	D	0	D

Consistency Check Operation:

line	A	B	C	D	E	F	G	H	Z
H at 0		0	$\overline{D}$	1	0	$\overline{D}$	D	0	D
E at 0	1	0	$\overline{D}$	1	0	$\overline{D}$	D	0	D

Thus, the test vector to detect line C s-a-1 fault is $ABCD = 1001$. Fortunately, in this example, we did not encounter an inconsistency. Note that we also had a choice to sensitize the path G2-G4-G5, instead of the path G2-G3-G5. If we sensitize the path G2-G4-G5, the test vector is $ABCD = 0101$.

In general, the D-algorithm is described in terms of *singular cover, Propagation D-cubes and D-intersection* [2]. However, a straightforward implementation such as given above is more economical in terms of memory requirement, and also faster in terms of processing time. The D-algorithm also guarantees to find a test vector if one exists. Many test-generation programs based on D-algorithm have been developed. However, for large circuits containing more than 10,000 gates, this algorithm is not economical. Another deficiency of this method is a large number of backtrack loops. The algorithm has no function to eliminate assignments that will obviously lead to an inconsistency. This problem is resolved in PODEM.

5.2.3 Algorithm PODEM

In PODEM, the test generation problem is considered as a branch and bound search of an n-dimensional binary state space (values at n primary inputs). In this algorithm, a decision tree is created in which two choices (1 and 0) are available at each decision node. Initially, a random choice is made. If an inconsistency occurs, algorithm backtrack the decision node to select another choice. In this algorithm, in the worst case, all possible input combinations are exhaustively examined as possible test patterns. This search is stopped as soon as a test vector is found. The main difference between PODEM and the D-algorithm is that PODEM uses a heuristic function in the backtrack operation. This heuristic consists of two major points:

1. Select the most easily controllable input if it alone can specify the output of the current gate, and thus allows us to obtain the current objective level. For example, a 0 for an AND/NAND gate and a 1 for an OR/NOR gate.
2. Select the least controllable input if the current objective level can only be achieved by setting all inputs to specify the output of the current gate.

The overall concept of PODEM is given in Figure 5.3 [3]. The algorithm is given in Figure 5.4. The procedures to determine the initial objective function and backtrack are given in Figures 5.5 and 5.6, respectively [3]. In PODEM, all primary inputs are initially considered as x, unknown value. The objective of the initial assignment is to put a D or $\overline{D}$ at the line on which a stuck-at fault is considered. This value is propagated to the output by assigning one value from the set [0,1] to a node, starting from unassigned primary inputs. If a conflict arises, the next value of the set [0,1] is assigned. If both values cause conflict, that node is removed from the decision tree and its child node is considered at x. This process continues until a test vector is found.

Example 5. Consider the fault, line G s-a-1, in the circuit given in Figure 5.7. To set a $\overline{D}$ at line G, the initial objective value is 0. This value is propagated backward to obtain an assignment $BCD = 111$. The implication of this assignment is $FH = 00$. Clearly, the fault cannot be propagated through the gate $G4$. To propagate the fault through $G5$, we assign line $I = 1$. Thus, the assignments at the primary inputs are $ABCDE = x1110$, which gives the required test vector.

PODEM is significantly better than the D-algorithm particularly for error correction and translation circuits. The PODEM-X (an IBM version) has been used for circuits containing up to 50,000 gates.

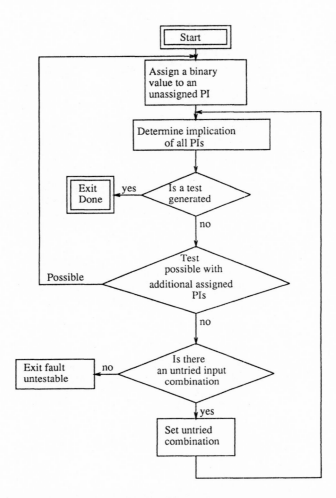

Figure 5.3 Overall concept of PODEM. (From [3], © 1981 IEEE. Reprinted with permission.)

5.2.4 Algorithm FAN

FAN is a modification of PODEM. It pays special attention to the fan-out points and uses multiple backtracks to reduce the test-generation time [4]. The major modifications are summarized as follows:

1. During each step of FAN; all values are traced forward and backward to determine the unique signal values. In other words, the implication operation is done after each step to know the unique values.

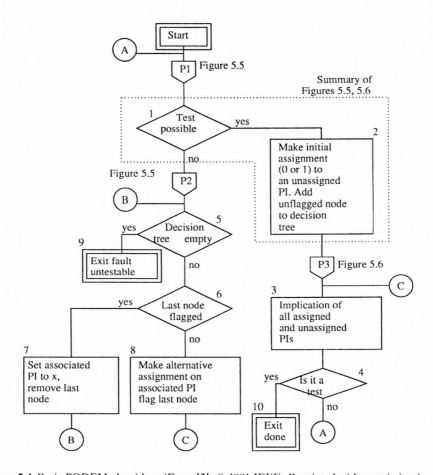

Figure 5.4 Basic PODEM algorithm. (From [3], © 1981 IEEE. Reprinted with permission.)

2. Assign D or $\overline{D}$ if it is uniquely determined by the fault. For example, for an AND gate, if the output is s-a-0, a D at the output and all inputs at 1 are unique, but, if the output is s-a-1, all inputs are left unspecified at x.
3. If all paths pass through a single gate, a partial sensitization can be done for this gate (called *unique sensitization*) to find the unique values at one or more lines.
4. A backtrack is not required to continue in that part of the circuit which does not have any fan-out point.
5. Multiple backtracking is done instead of using a single path.

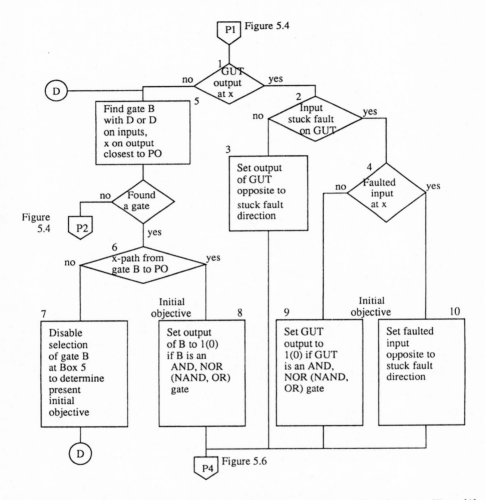

Figure 5.5 Determination of initial objective in PODEM to choose next PI assignment. (From [3], © 1981 IEEE. Reprinted with permission.)

The flowchart of FAN is given in Figure 5.8, with multiple backtracking in Figure 5.9 and functions to determine the final objective value in Figure 5.10.

The efficiency of FAN can be illustrated with a simple example. Consider the fault line L s-a-1 in the circuit given in Figure 5.11. For this fault, line L has a unique value $\overline{D}$. Lines J, K, and E also have unique values 1. A simple implication operation after assignment $JKE = 111$ gives the test vector $ABCE = 1101$ without any backtrack.

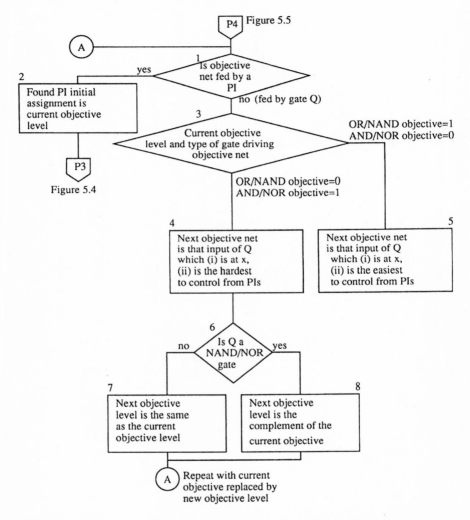

Figure 5.6 Backtrack procedure of PODEM. (From [3], © 1981 IEEE. Reprinted with permission.)

5.3 FAULT COVERAGE BY A TEST

The methods given in Section 5.2 generate tests by considering one fault at a time. Each test vector sensitizes a path from the primary inputs to the primary outputs. In general, each test vector covers more than one fault. For example, a test vector that detects an s-a-1 fault at the output of an OR gate also detects s-a-1 faults on

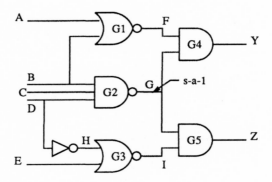

Figure 5.7 Combinational circuit for Example 5.

all of its inputs. If the test set is developed, considering each fault separately, it will contain many redundant vectors.

One possible solution is to re-examine the whole test set and delete the multiple entries of a test vector. This requires an additional exhaustive table look-up procedure and results in excessive cost of processor time.

Another solution is to use fault equivalence and dominance techniques, as described in Chapter 2, to collapse a fault set. For example, all equivalent faults can be collapsed into a single fault. A set of collapsed faults truly represents all single faults in the circuit. Hence, a test set that covers these collapsed faults also covers all single faults in the circuit. This avoids multiple entry of a vector in a test set. However, this procedure is also expensive in terms of CPU cost as it needs a fault simulator.

To generate test vectors, consideration of each fault separately is not necessary. An algorithm called *critical path tracing* allows us to consider fault equivalence during test vector generation.

5.3.1 Critical Path Tracing

Critical path tracing is an approximate algorithm that obtains test vectors by simulating a fault-free circuit. It traces a path backward from a primary output to the primary inputs. During this tracing process, it identifies all the critical values. The value obtained at the input is a test for all lines that contain a critical value [5]. The critical value is defined as follows.

Critical Value: A line is called to have a critical value c in a test if the test detects a stuck-at-$\bar{c}$ fault on that line.

84

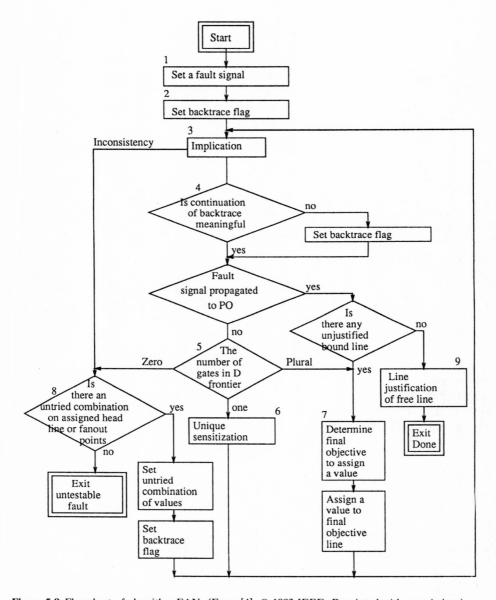

Figure 5.8 Flowchart of algorithm FAN. (From [4], © 1983 IEEE. Reprinted with permission.)

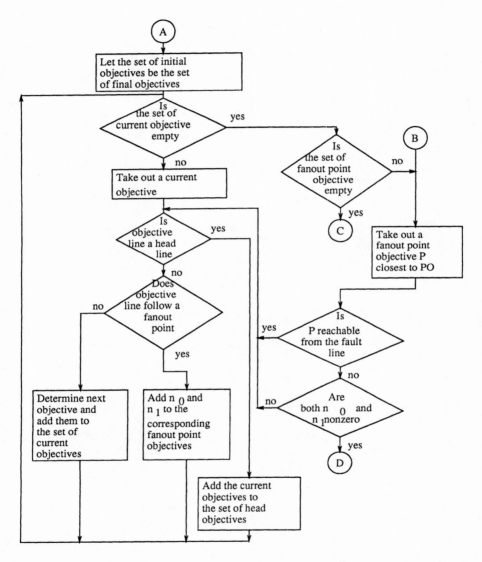

Figure 5.9 Multiple backtrack in FAN. (From [4], © 1983 IEEE. Reprinted with permission.)

Sensitivity: An input to a gate is called sensitive if the complementary value on that input causes a change at the gate output. For example, an input with 0 is sensitive for a NAND/AND gate.

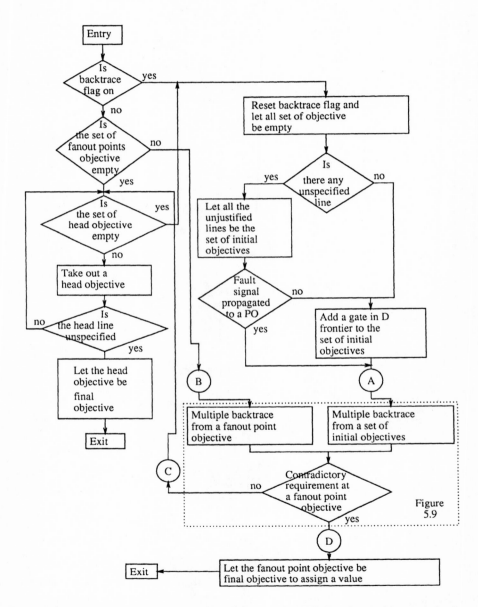

Figure 5.10 Procedure for final objective. (From [4], © 1983 IEEE. Reprinted with permission.)

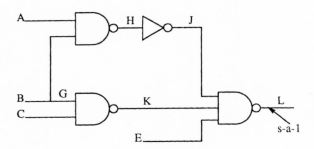

Figure 5.11 An example combinational circuit to illustrate the efficiency of FAN. (From [4], © 1983 IEEE. Reprinted with permission.)

Example 6. Consider the circuit given in Figure 5.12(a). To start with output Z, we have two choices, i.e., assign Z as 0^c, or Z as 1^c. Let's assume we assign Z as 0^c. This implies that lines F, G, and H are all 0^c. These values are further traced and a possibility is marked in Figure 5.12a. Thus, a test vector $ABCD = 0xx0$ will detect line A s-a-1, line E s-a-1, line D s-a-1, line F s-a-1, line G s-a-1, line H s-a-1, and line Z s-a-1 faults. One can make a different choice for the critical line while tracing a path. Thus, a different test vector can be obtained, which will have different fault coverage.

If we assign output Z as 1^c, a possible assignment can be given as shown in Figure 5.12(b). Thus, a test vector $ABCD = 111x$ will detect line A s-a-0, line B s-a-0, line C s-a-0, line E s-a-0, line F s-a-0, and line Z s-a-0 faults. A test set to cover all stuck-at faults in the circuit of Figure 5.12 is given in Table 5.2 with associated critical lines.

Critical path tracing is not a deterministic method. Unlike the methods given in Section 5.2, it does not guarantee finding a test. However, in practice, there are only a few situations in which it fails to find a test. The major advantage of this method is that there is less work required for fault enumeration, fault collapsing, and fault simulation. It identifies the fault coverage during the test generation process.

5.3.2 Multiple Faults

Note that the coverage of many faults by one vector does not mean the detection of multiple faults. A multiple fault means simultaneous presence of many faults in a circuit. While the number of single stuck-at faults in a circuit with n lines is only $2n$, the total number of single and multiple faults is $3^n - 1$. The detection of multiple faults is a much more complex problem due to the fault-masking effect.

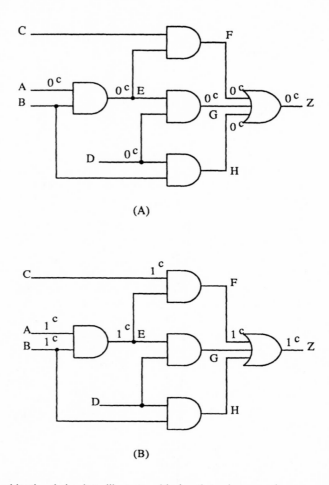

Figure 5.12 Combinational circuit to illustrate critical path tracing procedure.

Significant work has been done to detect multiple faults in a VLSI circuit. From the reliability point of view, a circuit should be tested to detect the presence of up to at least six faults. In a restricted type of circuit, a *single stuck-at fault test set* (SSFTS) also detects the majority of multiple faults. For example, in a fan-out-free circuit, SSFTS also detects all double and triple faults [6]. In [7], a method is presented to quantify the coverage of multiple faults by SSFTS. In an internal fan-out-free circuit, SSFTS also detects 98% of multiple faults of size six or less. For reconvergent fan-out circuits, first fan-out masks approximately 5% of the faults. The second and third fan-out masks an additional 3% of the faults in each case.

Table 5.2
A test set obtained from critical path tracing to cover all line stuck-at faults
in the circuit of Figure 5.12.

Test Vector	Line Status								
ABCD	A	B	C	D	E	F	G	H	Z
x0xx		0^r			0^r	0^r	0^r	0^r	0^r
0xxx	0^r				0^r	0^r	0^r	0^r	0^r
xx00			0^r	0^r		0^r	0^r	0^r	0^r
111x	1^c	1^c	1^c		1^c	1^c			1^c
11x1	1^c	1^c		1^c	1^c		1^c		1^c
x1x1		1^c		1^c				1^c	1^c

In some specific circuits, coverage of multiple faults may be as low as 76% with three fan-outs. In general, due to the cost of test vector generation, only single stuck-at faults are considered. After development SSFTS, fault simulation determines the exact fault coverage.

5.4 RANDOM TEST GENERATION

The test generation methods given in Sections 5.2 and 5.3 are deterministic. Also possible is to test a digital circuit by applying random inputs. This method of testing is called *random testing*. A random input pattern is applied to the CUT and also to a known fault-free circuit (called *gold unit*). The fault is detected by comparing the responses. Fault-free response for comparison can also be obtained by simulation. The concept of random testing is illustrated in Figure 5.13.

A widely used technique of random testing estimates the probability of sensitizing a fault. It is equivalent to calculating the probability of a 1 (or 0) on a particular line, while inputs to the circuit are random. By using this probabilistic measure, the length of a random test sequence is determined to obtain the desired fault coverage [8].

Let probabilities of 1 and 0 at the output of a gate at level k of the circuit be expressed as P_k^1 and P_k^0, respectively. Then, P_k^1 and P_k^0 for various gates can be given as follows.

For an n-input NAND gate:

$$P_k^0 = (P_{k-1}^1)^n = (1 - P_{k-1}^0)^n \tag{5.6a}$$

$$P_k^1 = 1 - P_k^0 = 1 - (P_{k-1}^1)^n = \sum_{i=1}^{n} P_{k-1}^0 \tag{5.6b}$$

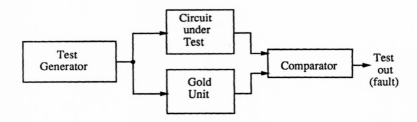

Figure 5.13 Pictorial view of random testing.

The probability P_k^0 is equal to the probability of n 1s at the $(k - 1)$ level. The P_k^1 is the probability of a 1 at level k. A 1 at level k will occur, if there is any 0 at level $(k - 1)$. The similar expressions for an n-input NOR gate are:

$$P_k^0 = 1 - P_k^1 = 1 - (P_{k-1}^0)^n = \sum_{i=1}^{n} P_{k-1}^1 \tag{5.7a}$$

$$P_k^1 = (P_{k-1}^0)^n = (1 - P_{k-1}^1)^n \tag{5.7b}$$

For an n-input AND gate:

$$P_k^0 = 1 - P_k^1 = 1 - (P_{k-1}^1)^n = \sum_{i=1}^{n} P_{k-1}^0 \tag{5.8a}$$

$$P_k^1 = (P_{k-1}^1)^n = (1 - P_{k-1}^0)^n \tag{5.8b}$$

For an n-input OR gate:

$$P_k^0 = (P_{k-1}^0)^n = (1 - P_{k-1}^1)^n \tag{5.9a}$$

$$P_k^1 = 1 - P_k^0 = 1 - (P_{k-1}^0)^n = \sum_{i=1}^{n} P_{k-1}^1 \tag{5.9b}$$

Using Equations (5.6) to (5.9), the probability of sensitizing a fault in a combinational circuit can be calculated.

Example 7. Consider the circuit given in Figure 5.14. Let us assume that the probabilities of a 1 on lines A and B are .5, and for lines C and D are .75. So, for line E, the probability of a 1 (P_E^1), is .25, and probability of a 0 (P_E^0) is .75. For line F, P_F^1 is .5625, and P_F^0 is .4375. Finally, $P_G^0 = P_F^0 + P_E^0$. This gives $P_G^0 = .328$, and $P_G^1 = .672$. These probabilities, in effect, tell us the possibility of detecting a stuck-at fault using random inputs.

To obtain a generalized relation, consider a K level NAND-NAND circuit. Let $P(K)$ be the probability of sensitizing a path of length K. The $P(K)$ represents the detection probability and is given as [8]:

$$P(K) = P(0) \prod_{i=0}^{K-1} (P_i^1)^{n-1} \tag{5.10}$$

Here $P(0)$ is the probability of occurrence of a D or $\overline{D}$ at the faulty line to sensitize the fault. For example, if we consider an input line stuck-at fault $P(0) = 1$. The term $(P_i^1)^{n-1}$ represents $(n-1)$ 1s at level i. If $P(K, M)$ is the probability that at least one vector out of M will sensitize a path of length L, $P(K, M)$ is given by:

$$P(K, M) = 1 - \text{(probability that none of the } M \text{ vectors sensitize the path)},$$

or

$$P(K, M) = [1 - P(K)]^M \tag{5.11}$$

Equation (5.11) can be solved to find a value of M, the minimum number of random vectors required to sensitize a path, i.e.,

$$M = \frac{\ln [1 - P(K, M)]}{\ln [1 - P(K)]} \tag{5.12}$$

For a fan-out-free K-level NAND-NAND circuit, Equation (5.12) is plotted for different detection probabilities. Figure 5.15 provides the confidence level and

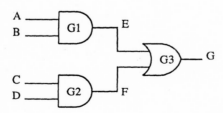

Figure 5.14 Circuit for Example 7.

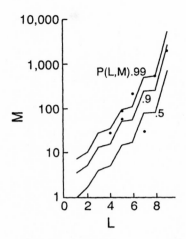

Figure 5.15 Number of random patterns required to sensitize a path at level 1 with 99%, 90%, and 50% probabilities. Average fan-in is 2. (From [8], © 1975 IEEE. Reprinted with permission.)

effectiveness of random testing. The nature of fault coverage by random input patterns, in general, is shown in Figure 5.16.

Another criterion to measure the effectiveness of random patterns is the coverage of hard to sensitize faults. If a circuit contains h hard to sensitize faults, with worst fault detection probability P, the minimum and maximum number of random patterns is given by [9]:

$$N_{upper\ bound} = \frac{\ln\left(\frac{e_t}{h}\right)}{\ln\ (1\ -\ P)} \tag{5.13}$$

$$N_{lower\ bound} = \frac{\ln\left(\frac{e_t}{h}\right)\ -\ \ln\left(1\ -\ \frac{e_t}{2}\right)}{\ln\ (1\ -\ P)} \tag{5.14}$$

Here, e_t is the minimum escape probability for a fault such that at least one fault of h will not be detected. A typical value for e_t is .001. Thus, $1\ -\ e_t$ is the confidence level that all faults will be detected. Note that the upper bound on the length of random patterns does not increase sharply with increasing number of faults, h. The rate of increase is logarithmic with respect to h. This implies that, while a moderate fault coverage can be obtained very efficiently with random

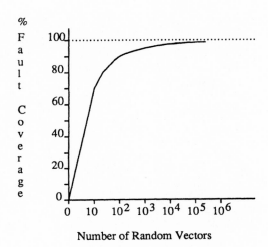

Figure 5.16 The nature of fault coverage by random testing.

patterns, a high fault coverage requires excessively large numbers of random patterns.

Generation of true random inputs is a costly process. To reduce the cost of random test generation, pseudorandom test vectors are used. An *n*-bit pseudorandom input can be obtained by an *n*-bit *linear feedback shift register* (LFSR). This technique is efficiently used as a built-in self-test and design for testability. The details of these methods are given in Chapter 10.

5.5   TEST GENERATION AT SWITCH LEVEL

Testing at the gate level is not adequate. As discussed in Chapters 3 and 4, many physical defects are manifested as bridging and open faults. A single stuck-at fault test set provides poor coverage of bridging and open faults. To obtain good fault coverage, one needs to test a circuit at the switch level. In recent years, some research effort has been directed toward test generation at the switch level. However, it is still an unresolved problem. An acceptable test generation procedure has not been developed at the switch level to date.

In current test-generation procedures at the switch level, a digital logic gate is represented as a connection graph. The nodes in the graph represent the electrical nodes, and the edges represent the transistors and are labeled by the signal to the particular transistor's gate. The nodes are connected by undirected edges representing a conduction path from the source to the drain. Each graph has two prime nodes representing either the V_{dd} and the output, or the GND and the output nodes of the gate. The conduction graph for p-part having prime nodes V_{dd} and output is called the p-graph, and the conduction graph for n-part with prime nodes GND and output is called the n-graph. The examples of such connection graphs for the NAND and the NOR gates are given in Figure 5.17.

To generate the test vectors, all possible paths from the V_{dd} to output in the p-graph, and from output to GND in the n-graph are identified. Each path is then examined for a shorted edge as well as a broken edge. If there are two or more paths in parallel, only one is sensitized. All other paths are kept off by applying a 0 to at least one edge. To test for a shorted edge in the n-graph (p-graph), a 0 (1)

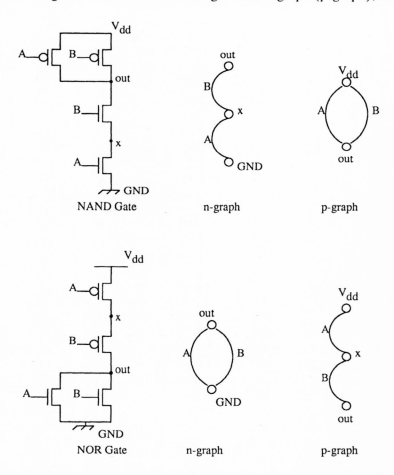

Figure 5.17 Schematic diagrams of two-input CMOS NAND and NOR gates with n- and p-graphs.

is applied to that edge while 1s (0s) are applied to all other edges in that path. To test for a broken or open edge in the n-graph (p-graph), a 1 (0) is applied to all the edges in that path. For example, the test sets for the NAND and the NOR gates of Figure 5.17 are given in Tables 5.3(a) and 5.3(b).

Test vectors for multi level combinational circuits can be generated by tracing the values from the graphs of last level to the first. If an edge requires a 1 (0), one path in the p-graph (or n-graph) of a previous stage is sensitized. This backward propagation continues until values at the primary inputs are assigned. The procedure is illustrated with an example.

Table 5.3(a)
Test vectors for the NAND gate; the test set is $AB = [01, 10, 11]$.

NAND Gate	Test AB	Test AB
p-graph	edge A short: 11	edge A broken: 01
	edge B short: 11	edge B broken: 10
n-graph	edge A short: 01	edge A broken: 11
	edge B short: 10	edge B broken: 11

Table 5.3(b)
Test vectors for the NOR gate; the test set is $AB = [01, 10, 00]$.

NOR Gate	Test AB	Test AB
p-graph	edge A short: 10	edge A broken: 00
	edge B short: 01	edge B broken: 00
n-graph	edge A short: 00	edge A broken: 10
	edge B short: 00	edge B broken: 01

Example 8. Consider the circuit given in Figure 5.18(a). The connection graphs for this circuit are given in Figure 5.18(b). In this circuit, G4 is an NOR gate at the last stage. From Table 5.3(b), we need LM [00, 01, 10]. Tracing these values backward, the following test vectors are obtained:

$$LM = 00 \rightarrow AB = 11, CD = [1x, x1]$$

$$D = x \rightarrow EF = xx; \text{ test is } ABCEF = 111xx$$

$$D = 1 \rightarrow EF = 00; \text{ test is } ABCEF = 11x00$$

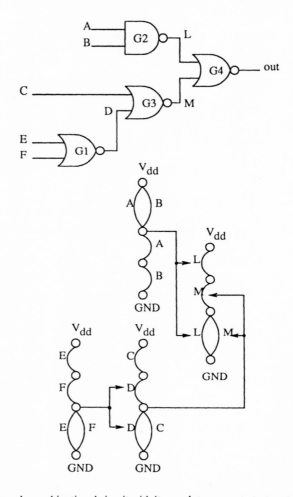

Figure 5.18 An example combinational circuit with its graph to generate test vectors at the switch level.

$$LM = 01 \rightarrow AB = 11, \; CD = 00$$

$$D = 0 \rightarrow EF = [1x, x1]; \; \text{test is } ABCEF = [1101x, 110x1]$$

$$LM = 10 \rightarrow AB = [0x, x0], \; CD = [1x, x1]$$

$$D = x \rightarrow EF = xx; \; \text{test is } ABCEF = [0x1xx, x01xx]$$

$$D = 1 \rightarrow EF = 00; \; \text{test is } ABCEF = [0xx00, x0x00]$$

Thus, a complete test set for Figure 5.18(a) is $ABCEF = [111xx, 11x00, 1101x,$ $110x1, 0x1xx, x01xx, 0xx00, x0x00]$.

Some important observations can be made here:

1. In a CMOS circuit, a test set which detects all single p-MOS stuck-on faults also detects all single line s-a-0 faults. A test set which detects all single n-MOS stuck-on faults also detects all single line s-a-1 faults.
2. It is sufficient to test a CMOS circuit for all transistor stuck-on faults to cover all line stuck-at faults.

The above procedure can generate test vectors very efficiently for all faults in n-MOS circuits. For CMOS circuits, due to charge retention at the output node as discussed in Chapter 4, this procedure is not sufficient to cover stuck-open faults. This procedure generates only the sensitizing vectors. Each vector generated by this method requires an initialization vector before it can be applied to the circuit. In addition, there are test invalidation problems, as discussed in Chapter 4. However, if the CMOS circuits are designed by using testable gates as given in Figure 4.7, the above test-generation procedure is quite efficient to generate test vectors at switch level.

5.6 FAULT SIMULATION

Before applying the test vectors to a circuit, the circuit is simulated under possible faulty conditions using a fault model. The objective of fault simulation is:

1. Grade a test set to determine its fault coverage; this evaluates the fault detection capability of a test set.
2. Determine a target fault that has not been detected by the present test set.
3. Build a fault dictionary if required; this also helps in the fault diagnosis process.
4. Obtain an estimation of circuit behavior in the presence of a fault.

Simulation, in general, is a very slow process. The cost of simulation is characterized in terms of processor time. Even for a moderately sized circuit, this cost may be significantly high. Assume a circuit with L lines and n inputs. Also assume that it takes t seconds to simulate one line under one fault for one vector. Thus, $L \cdot t$ seconds are required to simulate a circuit under one vector for one fault. As there are $2L$ stuck-at faults, we need $2L^2 \cdot t$ seconds to simulate a circuit under one vector. This gives $2^{n+1} \cdot L^2 \cdot t$ as an approximate simulation time. For a moderate size circuit ($n = 20$ and $L = 100$), we need about 5.8 hours just to simulate it for stuck-at faults if t is 1 μsec. Various fault simulation methods have been developed to reduce this large processor time.

Two popular methods to simulate a logic circuit are (i) *compiler-driven simulation* and (ii) *event-driven simulation*. In compiler-driven simulation, a circuit is translated into a machine-executable code. This code is executed under various input combinations to obtain information about the circuit output. In this process, the whole circuit is analyzed for each input.

In digital circuits, in general, only a fraction of all lines are active at a given time. In event-driven simulation, a gate is simulated only if an input to this gate is changed. This results in a significant reduction in computation time compared to the compiler-driven simulation.

The most widely used fault simulation methods are (i) *parallel fault simulation*, (ii) *deductive fault simulation*, and (iii) *concurrent fault simulation* [10]. All of these methods can be implemented in compiler-driven or event-driven environments.

In parallel fault simulation, a circuit is simulated by injecting multiple faults. It can be viewed as a pipeline process. A data word (generally a multiple of eight-bits) is used, in which each bit represents a different fault condition. Thus, by using a byte, a fault-free circuit can be simulated for seven different fault conditions. An important consideration is that the faults be injected such that the individual fault affects only a single bit. In this method, both fault-free and faulty circuits are simulated under all inputs, even if there is no disagreement at the output.

Deductive fault simulation is based upon the fact that, if the inputs to a circuit and the logical fault effect for these inputs are known, the circuit output can be calculated. A fault list is created for each line in the circuit. This fault list details every fault on the line by a separate name, and can be stored as a link list. This fault list propagates through the circuit by calculating the values at different lines by simple deduction.

Concurrent fault simulation is very similar to deductive fault simulation. In this method, an extensive fault list is maintained, which contains every fault originated on the line, as well as the fault effect on it due to some other fault in the circuit. This increases the size of the fault list as compared to that in deductive fault simulation. However, it requires less processor time as only those components are simulated with fault-free and faulty outputs that are different.

5.7 SUMMARY

Exhaustive testing of a combinational circuit is practically impossible. A single line stuck-at fault model is widely used to generate test vectors for combinational circuits. In general, a gate-level description is used to generate test vectors.

We have discussed Boolean difference, single-path sensitization, D-algorithm, PODEM, and FAN algorithm to generate test vectors for single line stuck-at faults.

While Boolean difference generates all possible tests, it is very time consuming. In path-sensitization methods, FAN is considered to be the most efficient. Although, in the test generation process, a test considers a single fault, each vector covers more than one fault. The test set generated by a straight algorithm without an optimization procedure contains redundancy. The fault collapsing technique is necessary to avoid multiple entry of a vector in the test set. The critical path tracing method is given and identifies the faults covered by a vector.

Random test generation and fault coverage have been discussed. The methods for calculating probabilistic fault coverage by random vectors are given. Finally, test generation at the switch level and fault simulation methods are discussed.

PROBLEMS

1. Obtain test vectors for lines A, C, F, and H for the circuit of Figure 5.2 using Boolean difference. Repeat with D-algorithm. Also, obtain tests for line H and G in Figure 5.1 using D-algorithm.
2. Compare D-algorithm, PODEM, and FAN in obtaining a test vector for the line G s-a-1 fault in Figure 5.12.
3. Obtain a complete test set using critical path tracing for Figures 5.1 and 5.2.
4. Let the probability of having a 0 or 1 at the primary input is same. What is the probability to sensitize the fault in Figure 5.2? What about line Z s-a-0 fault?
5. Obtain a complete test set using switch level approach as given in Section 5.5 for Figure 5.1. Compare it with Problem 5.3.

REFERENCES

1. F.F. Sellers, M.Y. Hsiao and C.L. Bearson, "Analyzing Errors with the Boolean Difference," *IEEE Trans. Computers*, **17**(7), July 1968, pp. 676–683.
2. J.P. Roth, "Diagnosis of Automata Failures: A Calculus and a Method," *IBM J. Research and Development*, **10**, July 1966, pp. 278–291.
3. P. Goel, "An Implicit Enumeration Algorithm to Generate Tests for Combinational Logic Circuits," *IEEE Trans. Computers*, **30**(3), March 1981, pp. 215–222.
4. H. Fujiwara and T. Shimono, "On the Acceleration of Test Generation Algorithms," *IEEE Trans. Computers*, **32**(12), Dec. 1983, pp. 1137–1144.
5. M. Abramovici, P.R. Menon, and D.T. Miller, "Critical Path Tracing–An Alternative to Fault Simulation," *Proc. Design Auto. Conf.*, 1983.
6. J.P. Hayes, "A NAND Model for Fault Diagnosis in Combinational Logic Networks," *IEEE Trans. Computers*, **20**(12), Dec. 1971, 1496–1506.
7. V.K. Agarwal and A.S. Fung, "Multiple Fault Testing of Logic Circuits by Single Fault Test Sets," *IEEE Trans. Computers*, **30**(11), Nov. 1981, pp. 855–864.

8. P. Agrawal and V.D. Agrawal, "Probabilistic Analysis of Random Test Generation Method for Irredundant Combinational Logic Networks," *IEEE Trans. Computers*, **24**(7), July 1975, pp. 691–695.

9. J. Savir and P.H. Bardell, "On Random Pattern Test Length," *IEEE Trans. Computers*, **33**(6), June 1984, pp. 467–474.

10. M.A. Breuer and A.D. Friedman, *Diagnosis and Reliable Design of Digital Systems*, Computer Science Press, Boulder, CO, 1976, pp. 224–242.

Chapter 6
Testing of Structured Designs (PLAs)

6.1 INTRODUCTION

Programmable logic arrays (PLAs) are widely used because of their cost effectiveness in implementing combinational switching functions. The testing of PLAs has proved to be a difficult problem [1]. The problem is due to the memorylike structure of the PLA and the large number of internal fan-ins and fan-outs. A large number of test vectors are required just to test for $(2n + m)p$ single-crosspoint faults, where n is the number of inputs, p is the number of product lines, and m is the number of outputs. Even for a small PLA, the test set becomes quite large. A system designer spends a significant amount of time on testing a programmed PLA before using it.

In the literature, several different PLA designs have been reported to simplify the PLA testing. These designs provide increased observability and controllability by accessing the internal nodes. Such designs are commonly known as *easily testable PLAs*. Easily testable PLA designs use some additional hardware to control and to observe each product and input line independently. A trade-off is found in the easily testable PLAs among the fault coverage, number of test patterns, functional independence, amount of extra hardware, and delay per test pattern. This chapter gives the basic concepts of testable PLA designs. Fault detection and diagnosis techniques for these designs are mentioned and methods to isolate a fault and reconfigure the PLA are given.

6.2 STRUCTURE OF A PLA

The PLA is a two-level implementation of a Boolean expression. Generally, a two-level NOR-NOR form is used to implement a function. The first-level NOR gates are called the *AND array* and the second level NOR gates are called the *OR array*.

Figure 6.1 shows a sample PLA with three inputs, three outputs, and four product lines. The PLA has been programmed to realize the following functions:

$$f1 = \bar{a}\,c + a\,\bar{b};$$

$$f2 = ab\,\bar{c} + b;$$

$$f3 = \bar{a}\,c.$$

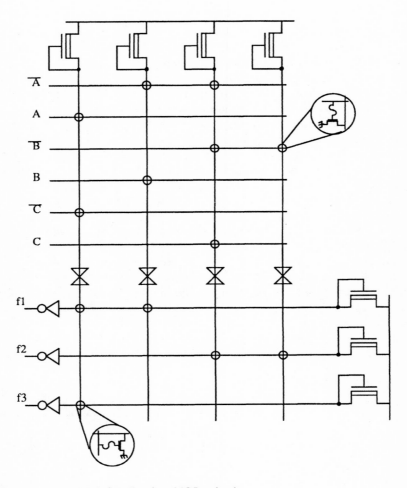

Figure 6.1 A sample NOR-NOR PLA in n-MOS technology.

Figure 6.1 also shows the structure of crosspoints. Every crosspoint consists of a transistor. In the AND plane, the gate of the transistor is connected to the input line through a switch that can be programmed to be on. In the OR plane, the gate of the transistor is connected to the product line through the switch. Figure 6.2 shows the structure of a semiconductor switch that is normally off, but can be programmed to be on. An intrinsic polysilicon film is sandwiched between two N^+ diffusion regions. The structure offers a resistance on the order of 10^9 Ω. When a high energy laser pulse is applied, the surface is damaged. Thus, the intrinsic part converts to the N^- type, offering a low resistance on the order of 10^3 Ω. These semiconductor switches are known as *antifuses*. Similarly, semiconductor fuses (thin-film Al wire) can be used at the crosspoints. Because of the presence of fuses (antifuses), PLAs suffer from different failure modes as described in Chapter 2.

There are several standard techniques to generate test vectors for the combinational circuits given in Chapter 5. If a PLA is modeled as a two-level NOR-NOR circuit, test vectors can be obtained with any of these methods. However, due to high internal fan-ins and reconvergent fan-outs, standard methods do not provide high fault coverage. In addition, standard techniques are restricted to stuck-at faults. The other failure modes, i.e., crosspoint faults, are not covered by the standard test algorithms. The situation is best illustrated with an example.

Example 1. Consider the PLA given in Figure 6.1. The two-level NOR-NOR model for this PLA is shown in Figure 6.3. All line stuck-at faults are listed in Table 6.1, which also contains test vectors generated by the D-algorithm.

One may observe that some line stuck-at faults are not detected in the equivalent gate-level model of Figure 6.3. Gate-level models do not need line $\overline{b}$ due to redundancy, and hence any stuck-at fault on this line is not detected. Again, due to redundancy, for output $f2$, there is no test vector for the product line #5 stuck-at fault. To sensitize a stuck-at fault on line #5, $\overline{b}$ should be 0. However, to

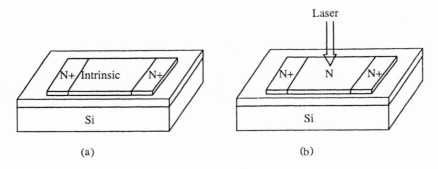

Figure 6.2 Structure of antifuse in PLA: (a) $R = G\Omega$; (b) $R = k\Omega$.

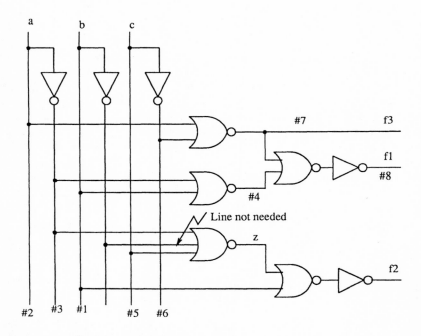

Figure 6.3 Gate-level model of example PLA given in Figure 6.1.

Table 6.1
Test vectors for the gate-level circuit given in Figure 6.3;
x represents a don't-care value.

Line Number	Faults s-a-1 ABC	Fault s-a-0 ABC
1	$x0x$	$x1x$
2	$0x1$	$1x1$
3	$10x$	000
4	$11x$	$10x$
5	—	—
6	$0x1$	$0x0$
7	$1xx$	$0x1$
8	$11x$	$10x$

propagate this fault effect, b should be 0. This contradictory assignment is not possible, and hence there is no test vector for line #5 in the gate-level model. Another inadequacy of the gate-level model becomes clear in the presence of a crosspoint fault, which changes the functionality of the PLA. Gate-level models fail to accommodate the change in functionality in the presence of a crosspoint fault.

The above example shows the inadequacy of standard test generation methods using gate-level models. Specialized methods are used to generate tests for PLAs. Techniques have been developed to verify the functionality of PLAs by utilizing their structural properties.

6.3 EASILY TESTABLE PLA

Due to difficulties in testing PLA with standard techniques, testable designs have been developed [1–5]. These designs allow independent control of the product lines and input lines to check for corresponding crosspoint faults. The original scheme as given in [2] and [3] tests a PLA independent of its function for all single-crosspoint faults and all single stuck-at faults. This scheme is modified by reducing additional hardware [1]. However, the modified schemes have either increased the number of test patterns, or made the test phase function-dependent [6]. Thus, the modified schemes have increased the computational complexity of test pattern generation. The total test time, which depends on the number of test patterns, delay per test, and computational complexity, is the same in each scheme.

6.3.1 PLA Testing with Parity Trees

The easily testable PLA designs use parity trees at the product lines and output lines to detect a fault. The testable design can be obtained by adding the following to the PLA:

1. *Pass transistors at the input lines.* A signal $c1$ controls all pass transistors associated with the true inputs while another signal, $c2$ controls all pass transistors associated with the complementary inputs. During normal operation, $c1 = c2 = 1$, keeping the pass transistors on. During the test mode $c1 = \overline{c2}$. This allows the activation of either the true inputs or the complementary inputs during testing.
2. *An extra column of transistors in the AND plane.* With this extra column, if the number of columns is even in each row in the AND plane, the number of absence of crosspoints is odd. Thus, in the absence of a fault, each input line will activate an odd number of product lines. If the number of columns is odd, then the number of absence of crosspoints on each row is even. In

this case, in the absence of a fault, an even number of product lines are active.

3. *An XOR tree at the product lines to check parity.* The output of this tree is referred to as *AND parity*.

4. *An extra row in the OR plane.* With this extra row, if the number of rows in the OR plane is odd in each column, an even number of absence of crosspoints are made. Thus, in the absence of a fault, an even number of output lines are active. If the number of rows is even, an odd number of absence of crosspoints are made on each column in the OR plane.

5. *An XOR tree at the output lines to check parity.* This is referred to as *OR parity*.

6. *A shift register connected to the product lines in the OR plane.* This register is used to apply the test patterns to the OR plane.

The augmented PLA, corresponding to the sample PLA of Figure 6.1, is given in Figure 6.4. XOR gates are shown in the cascade form. Alternatively, an XOR tree may be used. For this testable PLA, a universal test set can be obtained which is independent of the function implemented by the PLA.

6.3.2 Universal Test Set for Easily Testable PLAs

To test the testable PLA, the test set is divided into two parts. One set of vectors tests the AND plane and the second set tests the OR plane. To test the AND plane, first the true inputs are activated by making $c1 = 1$ and $c2 = 0$. One initialization vector is also required, which is the all 1 vector with $c1 = c2 = 1$. This vector ensures the absence of charge in the stray capacitances associated with the input lines. Test vectors follow this initialization vector. If the jth input is denoted by I_j, the tests are given by the n-tuple:

$$T = \{T_1, T_2, \ldots, T_i, \ldots T_n\}$$

where n is the number of test vectors. The ith test vector T_i ($i = 1, \ldots n$) is given by

$$T_i = [I_1 \cdots I_j \cdots I_n]$$

where

$$I_j = 1 \quad \text{if } i = j$$

$$I_j = 0 \quad \text{if } i \neq j$$

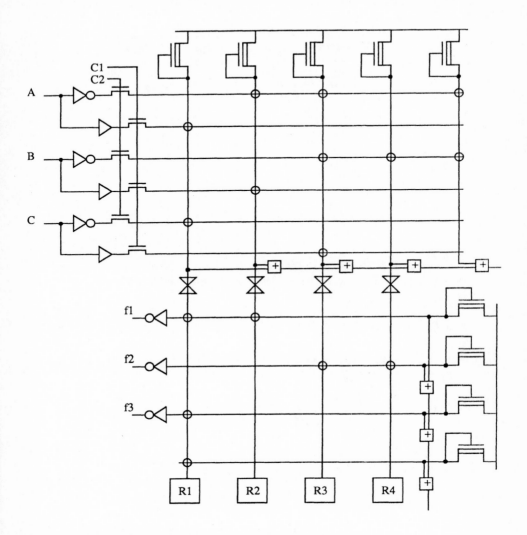

Figure 6.4 Augmented PLA with parity circuits for easy testing.

Each test vector basically sets one input to 1 and all other inputs to 0. The number of absence of crosspoints is odd (or even) on each input line, and an odd (or even) number of product lines is activated in the AND plane. Thus, in the absence of a fault, AND parity is odd (or even).

Similarly, the complementary inputs are activated, keeping the true inputs in the high impedance state. In this case, the initialization vector is $c1 = c2 = 1$, and all the inputs are 0s. The tests for the complementary inputs are given by

$$c1 = 0; c2 = 1; \text{ and } T = \{T_1, T_2, \ldots T_i \cdots T_n\}$$

The ith test vector is given by

$$T_i = [I_1 \cdots I_j \cdots I_n]$$

where

$$I_j = 0 \quad \text{if } i = j$$

$$I_j = 1 \quad \text{if } i \neq j$$

The test set for the OR plane is found in a similar fashion. If the ith product line in the OR plane is denoted by R_i, the test vectors for the OR plane are given by the p-tuple:

$$T = \{T_1 \cdots T_j \cdots T_p\}$$

where p is the number of product lines. The ith vector T_i $(i = 1, \ldots, p)$ is given by

$$T_i = [R_1 \cdots R_j \cdots R_p]$$

where

$$R_j = 1 \quad \text{if } i = j$$

$$R_j = 0 \quad \text{if } i \neq j$$

Each such vector sets one product line to 1 and all the other product lines to 0 in the OR plane. This in effect, activates an odd (or even) number of output lines. Thus, in the absence of a fault, OR parity is odd (or even).

The complete test set for the PLA given in Figure 6.4 is shown in Table 6.2(a) and 6.2(b). This test set not only detects all single line stuck-at, bridging, and crosspoint faults, but it also detects a majority of multiple faults.

The observations are summarized as follows:

1. In the testable PLA, all single crosspoint faults, all single stuck-at faults, and all single bridging faults between product lines associated with the AND array as well as the OR array and additional logic are detectable.

Table 6.2(a)
Test vectors for the AND plane.

c2	c1	A	B	C
0	1	1	1	1
0	1	1	0	0
0	1	0	1	0
0	1	0	0	1
1	0	0	0	0
1	0	0	1	1
1	0	1	0	1
1	0	1	1	0

Table 6.2(b)
Test vectors for the OR plane.

R1	R2	R3	R4
1	0	0	0
0	1	0	0
0	0	1	0
0	0	0	1

2. In the testable PLA, all single crosspoint faults, all single stuck faults, and all single bridging faults are detectable by a universal test set of length $[2(n + 1) + p]$. This is a minimal test set.
3. In the testable PLA, all odd number of crosspoint faults and line stuck-at faults are detectable by a universal test set designed to detect all single faults.

An even number of faults on a row in the AND plane or column in the OR plane cannot be detected in the augmented PLA by a universal test set. Multiple bridging faults in the AND plane, which involve more than one input line, are also not detectable.

6.3.3 Variations of Parity-Based Testable Design

Many variations are available in the PLA design given above. The PLA given in Figure 6.4 requires two parity trees, one register, two controls, one extra row, and one extra column. Instead of using two parity trees and one register, one can use two registers and one parity tree [4]. One register is used at the input and the second register is used at the product lines. The parity tree is used at the output lines. The schematic design is shown in Figure 6.5. Note that, in such a variation of design, the test generation procedure also changes slightly. However, the test set still can be universal, independently of the PLA's functionality.

Another variation in PLA design given in Section 6.3.1 is possible for reducing the total test time. By using a universal test set, test generation time is minimized.

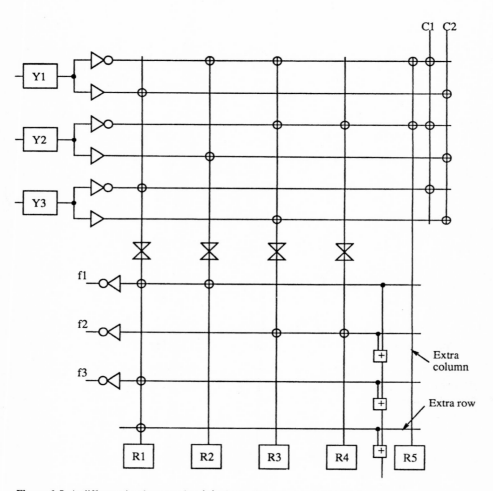

Figure 6.5 A different implementation [4] of a parity-based testable PLA.

Improvements are possible in test application time and response evaluation time. This improvement is obtained by simultaneously testing the AND and OR planes in parallel.

A control signal $c3$ is added to the buffers at each product line in the PLA design given in Figure 6.4 [6]. This signal is controlled such that the buffers present a high impedance state when $c3 = 0$. In normal operation, $c3 = 1$, providing transparent buffers. During testing, the AND and OR planes are isolated by making $c3 = 0$ and tested in parallel. This, in effect, reduces the total test time by a factor of almost 2, independently of the delay per test.

6.4 BUILT-IN SELF-TEST PLA

The PLA designs given in Section 6.3 require an external test pattern generator. Daehn and Mucha [7] suggested a design for on-chip test generation and response evaluation. Such methods, in general, are known as *built-in self test* (BIST) and will be discussed in detail in Chapter 10. In this design, three *built-in logic block observer* (BILBO) registers are used. One register is attached to the input lines, one register to the product lines, and one register to the output lines. The design is shown in Figure 6.6. Each BILBO register is designed to perform one of four possible functions:

1. Work as a shift register;
2. Work as a *linear feedback shift register* (LFSR) and a pseudorandom test generator;
3. Work as a signature compressor;
4. Work as multiple independent latches.

To test a PLA, BILBO #1 generates the test for the OR plane and BILBO #2 evaluates its response. BILBO #2 generates the test for the AND plane, while the response is evaluated in BILBO #3. The output inverters and input decoders are tested by the vectors generated by BILBO #3, while the responses are evaluated in BILBO #1. With this technique, all single stuck-at faults, bridging, and cross-point faults in the PLA are detectable. Note that, although this scheme eliminates the requirement of an external test pattern generator and response evaluator, it is extremely costly in terms of hardware overhead.

6.5 TESTING OF EEPLA

In recent years, EEPLAs have been developed to overcome the problem of irreversible programming. In EEPLAs, double polysilicon gate transistors are used instead of antifuses. These transistors work as reprogrammable switches at the crosspoints. The EEPLA structure and cell design are shown in Figure 6.7.

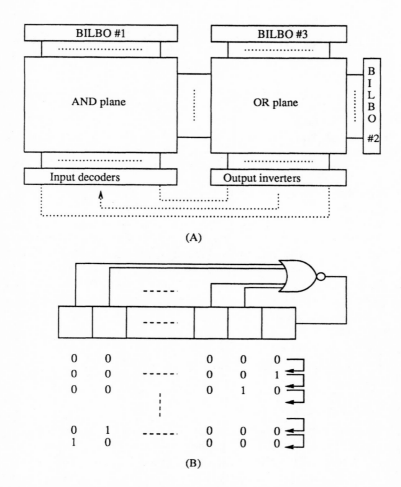

Figure 6.6 (a) Schematic design of BIST PLA; (b) test generation by BILBO.

The EEPLAs can be made testable by adding a single register to the product lines. This register is referred to as the *product line register* (PL register). This register allows an observer to check the response of the AND plane. Response of the OR plane is observed through the output register. This design is given in Figure 6.8, where all true and complementary signals are assumed to be independently controlled. This can be achieved by a minor modification in the input decoder. Note that the extra hardware used in this design is only a *p*-bit wide PL register. The test set for this EEPLA can be obtained as follows.

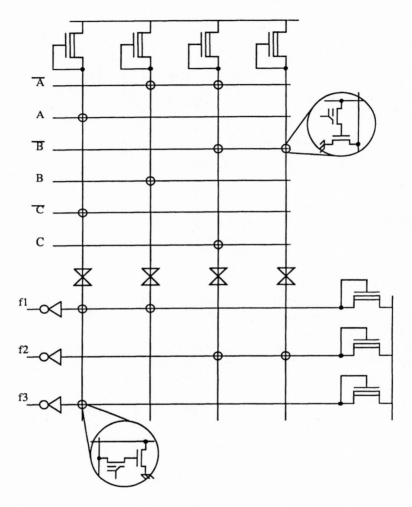

Figure 6.7(a) An example EEPLA in NOR-NOR form.

During testing, only one input is activated at a time (set to 1); all other inputs are set to 0. The AND plane is programmed such that all the inputs which are set to 0 have crosspoints at every product line. The input set to 1 has crosspoints at all product lines except one. Hence, under fault-free conditions, one product line is 1 and all others are 0. Consequently, only one bit in the PL register is 1, while all others are 0.

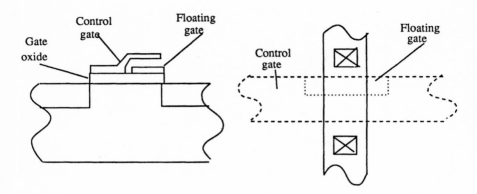

Figure 6.7(b) Schematic and layout of the double polysilicon gate transistor used in an EEPLA.

The OR plane is programmed similarly to the AND plane. Note that the AND plane is programmed such that only one product line sets to 1. For the OR plane, product lines behave as inputs. Hence, the product line that will assume logic 1 has crosspoint at every output line except one. All other product lines have crosspoints at every output line. Thus, under fault-free conditions, one output line is 1 and all others are 0. Therefore, only one bit in the output register is 1, while all others are 0. This EEPLA programming procedure for testing is given in Figure 6.9.

The test set for example EEPLA of Figure 6.8 is shown in Figure 6.10, where the AND plane and OR plane are observed separately, but simultaneously, in two different registers. The observations regarding testing of EEPLAs are summarized below:

1. In the testable EEPLA, all single crosspoint faults, all single product line stuck-at faults, and all single bridging faults between the product lines or output lines are detectable and can be diagnosed.
2. All single output line stuck-at faults and all single stuck-at faults in the output register are detectable, but cannot be distinguished from one another.
3. To test an EEPLA for all single crosspoint faults, all single stuck-at faults, and all single bridging faults, a test set of length np (if $n \leq m$), or mp (if $m \leq n$) is required. This is a minimal test set.

Note that in the testable EEPLA design, as such, the output register stuck-at faults and output lines stuck-at faults are not distinguishable from one another. However, if the output register is designed by using scan flip-flops, this distinction can be obtained by additional testing of the output register.

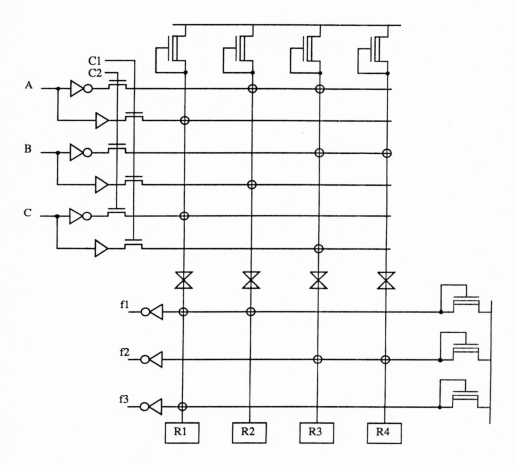

Figure 6.8 Augmented EEPLA for easy testing.

In EEPLAs, the program verification is also an important issue for the user. By using a bidirectional register instead of a simple PL register, program verification can be done in $(n + p)$ vectors.

To verify the programming of the AND plane, one input $(I_i; i = 1$ to $n)$ is set to 1, while all others are 0. In this condition, only those product lines which have crosspoint on input I_i go to 0. As the response is observed through PL register, an observer is able to verify the crosspoints at input I_i.

After verification of the AND plane, the OR plane can be verified, the product lines behave as inputs, and the response is observed through the output register. To verify the OR plane, one bit $(P_i; i = 1$ to $p)$ in the PL register is set to 1, while

```
/* Inputs are represented as Iᵢ; i = 1 to n
/* Product lines are represented as Pⱼ; j = 1 to p
/* Output lines are represented as Oₗ; l = 1 to m
/* Values at the inputs are defined as:
   For i = 1 to n Do
     Begin
       Iq = 1 if q = i
       Iq = 0 if q ≠ i
     end
/* Programming of the AND plane is defined as:
   For i = 1 to n Do
     Begin
       For j = 1 to p Do
         Begin
           no crosspoint at Ciw if w = j
           crosspoint at Ciw if w ≠ j
         end
       crosspoint at Cqj if q ≠ i
     end
/* Programming of the OR plane is defined as:
   For j = 1 to p Do
     Begin
       For 1 = 1 to m Do
         Begin
           no crosspoint at Cjr if r = 1
           crosspoint at Cjr if r ≠ 1
         end
       crosspoint at Csl if s ≠ j
     end
```

Figure 6.9 Programming procedure of EEPLA for testing.

all others are 0. Due to the bidirectional PL register, all product lines become 0 except P_i. In this condition, only those output lines which have crosspoints on P_i become 0. As the functionality is known, an observer can verify the OR plane by looking at the response through the output register.

6.6 TESTING FOR MULTIPLE FAULTS IN PLA

In an actual circuit, a processing defect causes multiple fault rather than a single fault. To analyze the testability of the proposed design for multiple faults, the concept of *fault equivalence* is used.

Inputs

		AND Plane		OR Plane		

$\overline{A}$	1	0111	1110	
A	0	1111	1111	
$\overline{B}$	0	1111 ------ 1111		
B	0	1111	1111	
$\overline{C}$	0	1111	1111	
C	0	1111	1111	

0111	1110	f1
1111 ------ 1111		f2
1111	1111	f3

←——————→
P vectors

←——————→
P vectors

0	1111	1111
1	0111	1110
0	1111 ------ 1111	
0	1111	1111
0	1111	1111
0	1111	1111

1111	1111
0111 ------ 1110	
1111	1111

0	1111	1111
0	1111	1111
1	0111 ------ 1110	
0	1111	1111
0	1111	1111
0	1111	1111

1111	1111
1111 ------ 1111	
0111	1110

←——————→
Total *MP*
combinations
($M = 3, P = 4$)

0	1111	1111
0	1111	1111
0	1111 ------ 1111	
1	0111	1110
0	1111	1111
0	1111	1111

←——————→
Total *NP*
combinations
($N = 6, P = 4$)

Figure 6.10 Test set for the EEPLA given in Figure 6.8. The length of the test set is *NP* as $N > M$.

Two faults are *equivalent* if and only if for each input the circuit response is identical under the faults.

Two faults are *subequivalent* if the circuit response is identical for at least one input vector under the faults.

In PLA design, the following observations can be made:

1. A bridging fault (single or multiple) is subequivalent to a multiple line stuck-at fault. It is also subequivalent to a multiple crosspoint fault.
2. A line stuck-at fault (single or multiple) is subequivalent to a multiple cross-point fault. It is also subequivalent to a multiple bridging fault.
3. A crosspoint fault (single or multiple), is subequivalent to a multiple bridging fault. It is also subequivalent to a multiple line stuck-at fault.
4. Any single or multiple fault consisting of bridging faults, line stuck-at faults or crosspoint faults is subequivalent to a multiple bridging fault. It is also subequivalent to a multiple line stuck-at fault as well as a multiple crosspoint fault.

In a PLA, most of the multiple faults are covered by a test set designed for single fault coverage. The argument for multiple fault coverage is developed as follows.

Suppose there is a multiple fault F in a PLA, which is a set of x simultaneous single faults, i.e.,

$$F = [f_1, f_2, \ldots, f_x]$$

A test vector that detects a single fault f_k will fail to detect a multiple fault F if the effect of single fault f_k is rectified by the other $(x - 1)$ faults represented as $(F - f_k)$. In other words, the multiple fault may remain undetected under a vector if the following condition is satisfied:

$$\text{Eff}(f_k) = \overline{\text{Eff}}(F - f_k)$$

where Eff represents the effect of a fault and $\overline{\text{Eff}}$ represents the complement of the effect. Although a test vector may fail, the multiple fault still may be detected while applying the whole test set. A multiple fault may only remain undetected if the effect of any single fault f_i ($i = 1$ to x) is rectified by the other $(x - 1)$ faults, represented as $(F - f_i$, for all $i = 1$ to x); at the same time, the effect of $(x - 1)$ faults is also rectified by the fault f_i. For this, the following conditions need to be satisfied:

$$\text{Eff}(f_i) = \overline{\text{Eff}}(F - f_i) \text{ and}$$

$$\text{Eff}(F - f_i) = \overline{\text{Eff}}(f_i); \text{ for all } i = 1 \text{ to } x$$

In general, for a PLA test set, a multiple crosspoint fault cannot mask the effect of a single crosspoint fault for all vectors. Hence, any multiple crosspoint fault is detectable. Similarly, a multiple stuck-at fault or multiple bridging fault cannot mask the effect of a single stuck-at fault or single bridging fault for all vectors. In fact, any multiple fault that is a set of several single faults of the same category is detectable by more than one vector. This is due to the fact that the faults of one category cannot mask the effect of a single fault in the same category for all vectors. Therefore, any multiple fault is detected by a test set that detects all single faults.

6.7 FAULT ISOLATION AND RECONFIGURATION

In recent years, considerable emphasis has been given to yield enhancement. Apart from the improved processing, research has been directed to isolate a fault and to use the fault-free part of the circuit. Generally, some extra lines are provided as redundant circuitry, which are used to reconfigure the PLA [10–11]. The faulty lines are identified and removed from the circuit with the help of a laser. The signal is rerouted so that the redundant lines can replace the corresponding circuitry.

Conceptually, these schemes appear to be simple. However, due to the use of a laser, fault isolation and replacement of the faulty circuit with the redundant circuit is a difficult problem. The routing of the signals after reconfiguration becomes significantly more complex.

The recent development of the double polysilicon gate transistor has simplified this problem. The solution is to use such a transistor on each input line, product line, and output line. After identifying a fault, faulty lines are easily removed from the circuit. Furthermore, redundant lines can be connected to replace the faulty lines with double polysilicon reprogrammable transistors without complicating the routing. In this scheme, redundant lines are used instead of the faulty lines, while the faulty lines are removed and remain unused.

The scheme is illustrated in Figure 6.11 for a testable EEPLA. The same concept is applicable for any PLA design. In the figure, one switch is provided on each input line, product line, and output line. These switches are kept on at the functional lines and off at the redundant lines. After testing, if a crosspoint fault is observed in the AND plane, the corresponding input line and product lines are removed by turning the switches off. One redundant input line and the product line is connected by turning the switches on. Similarly, in the case of a crosspoint fault in the OR plane, two lines (one product line and one output line) are replaced. In the case of a stuck-at fault, the corresponding line is replaced and, in the case of a bridging fault, two lines are replaced.

The number of extra lines determines how many faults can be tolerated. For example, two input lines, four product lines, and two output lines can tolerate any

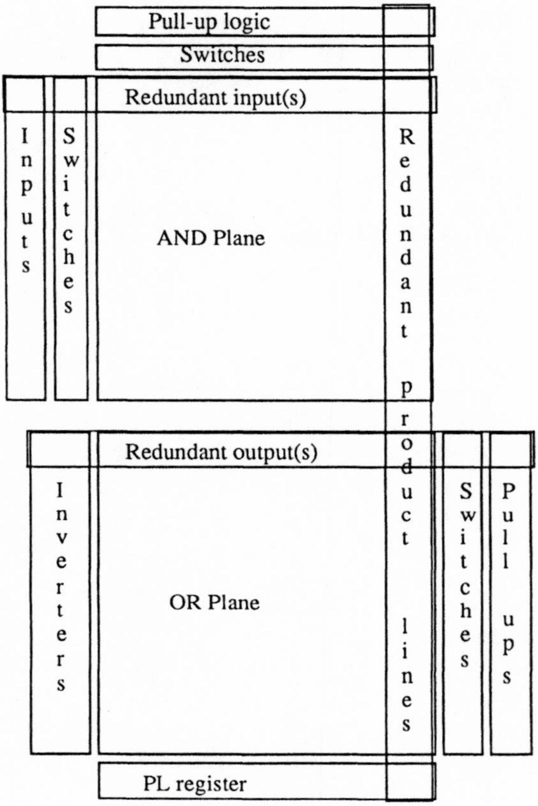

Figure 6.11 Schematic testable EEPLA for fault tolerance.

four crosspoint faults or any two bridging faults, one in the AND plane and the other in the OR plane. Analysis regarding fault tolerance versus redundant lines is given in Figure 6.12 [11].

6.8 SUMMARY

In this chapter, techniques are given for testing a PLA. PLA faults are examined and compared with faults in random logic. Test generation methods for random combinational logic are shown to be inadequate for PLAs.

Easily testable PLA designs are given. Parity trees at the product lines and the output lines can effectively test a PLA. The variations in testable PLA designs

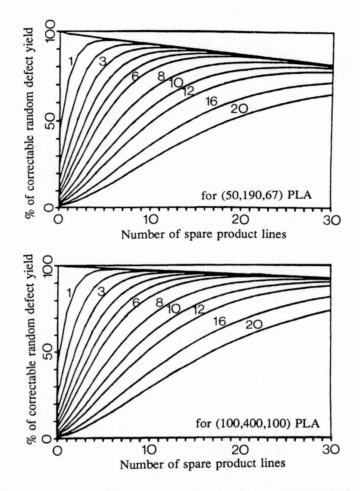

Figure 6.12 Trade-off in number of extra product lines and PLA yield. (From [11], ©1988 IEEE. Reprinted with permission.)

are discussed. The concepts of built-in self test PLA and the procedure to test EEPLAs are given. The various trade-offs and penalties due to hardware overhead are discussed.

In addition to fault detection, PLA fault diagnostic methods are given. By using reprogrammable switches, the PLA can be easily reconfigured. A scheme is given for isolating a fault and replacing the faulty circuit with redundant lines.

PROBLEMS

1. Make a PLA for adding two four-bit numbers.
2. Consider the PLA given in Figure 6.1, calculate the total number of possible single faults. How many faults can be modeled at the stuck-at level?
3. Prove that the test method given in Section 6.3 is indeed capable of detecting all single faults in a PLA. Why are not all multiple faults covered with this method?
4. Modify the PLA of Problem 1, according to Section 6.3. Obtain a complete test set for this modified PLA.
5. Consider the PLA of Problem 1 to be an EEPLA. Modify it to make it easily testable. Obtain a universal test set to test it completely.

REFERENCES

1. V.K. Agarwal, "Easily Testable PLA Design," Ch. 3 in *VLSI testing*, T. W. Williams, ed., North Holland, Amsterdam, 1986.
2. H. Fujiwara and K. Kinoshita, "A Design of Programmable Logic Arrays with Universal Tests," *IEEE Trans. Comp.*, **30**(11), Nov. 1981, pp. 823–828.
3. S.J. Hong and D.L. Ostapko, "FITPLA: A Programmable Logic Array for Function Independent Testing," *Proc. 10th Int. Symp. Fault Tol. Comp.*, 1980, pp. 131–136.
4. R. Treuer, V.K. Agarwal, and H. Fujiwara, "A New Built-in Self Test Design for PLAs with High Fault Coverage and Low Overhead," *IEEE Trans. Comp.*, **36**(3), March 1987, pp. 369–373.
5. XiAn Zhu and M.A. Breuer, "Analysis of Testable PLA Designs," *IEEE Design and Test*, Aug. 1988, pp. 14–28.
6. R. Rajsuman, Y.K. Malaiya, and A.P. Jayasumana, "A New Testable Design of Field Programmable Logic Arrays," *Proc. Int. Symp. Circuits and Systems (ISCAS)*, 1990, pp. 436–439.
7. W. Daehn and J. Mucha, "A Hardware Approach to Self Testing of Large Programmable Logic Arrays," *IEEE Trans. Comp.*, **30**(11), Nov. 1981, pp. 829–833.
8. R. Rajsuman, "Design of Reprogrammable FPLA," *Electronics Lett.*, **25**(11), May 1989, pp. 715–716.
9. R. Rajsuman, "Design of a Highly Testable and Fault Diagnosable, Reprogrammable FPLA," Technical Report CES-90-17, Department of Computer Engineering and Science, Case Western Reserve University, June 1990.
10. N. Wehn, M. Glesner, K. Caesar, P. Mann and A. Roth, "A Defect Tolerant and Fully Testable PLA," *Proc. Design Auto. Conf.*, 1988, pp. 22–27.
11. C.L. Wey, "On Yield Considerations for the Design of Redundant Programmable Logic Arrays," *IEEE Trans. CAD*, **7**(4), April 1988, pp. 528–535.

Chapter 7
Testing of Random Access Memory

7.1 INTRODUCTION

Memory is a very important part of a computer system. A significant amount of work has been done in recent years to obtain fast and very large memory systems. As a result, the density of semiconductor memory chips has increased dramatically. With the increasing complexity, the efficient testing of such memories has been recognized as a difficult problem. Because the complexity of memories is quadrupling every two to three years, even a linear increase in testing time becomes undesirable for large memories. A multimegabit random access memory requires excessive time just to test all cell stuck-at faults. To overcome this problem, researchers have sought to develop innovative test generation algorithms [1–5] and on-chip built-in test methods [6–10].

Because of the large testing time requirements, memory testing problems cannot be solved by clever test algorithms alone. In this chapter, we will first discuss test generation algorithms to obtain a minimal number of test vectors while covering 100% faults under the specified fault model. Second, to obtain a smaller testing time, we will discuss and evaluate various testable design methods to build (design) large memories using small blocks. Finally, fault diagnosis and memory reconfiguration schemes for memories are given.

7.2 TEST ALGORITHMS

Several innovative test algorithms for random access memories have been reported. These algorithms can be categorized into two classes. In one set of algorithms, the whole memory is read after changing the value of one or more cells. In the second class of test algorithms, only that cell with its value changed is read. In the literature, test algorithms have also been classified according to the fault model, i.e., algorithms based upon the cell stuck-at fault model and the pattern-sensitive fault model. In this section, we will describe the most widely used algorithms.

7.2.1 Algorithm GALPAT

This algorithm is also known as the *galloping 1* or *ping-pong* test. This algorithm first initializes all memory cells to 0. Then, for each cell, it writes 1, reads all the cells, and then writes back 0 to that cell. The procedure looks like a 1 is shifting its position from cell to cell throughout the whole memory. The algorithm is given in Figure 7.1. Note that a complementary pattern can also be used, i.e., instead of a 1, a 0 may be shifted through the memory.

In this algorithm, each cell is read at least once, when it has a value 0 as well as when it has a value 1. Thus, all cell stuck-at-1/0 faults are detected. The switching of state 0-to-1 and 1-to-0 also detects state transition faults. As only one cell contains 1 at any time and all cells are read, all address decoder faults are detected. Any two arbitrary cells under states 00, 10, and 01 are read, which covers a majority of state coupling faults. However, all state coupling faults are not covered. Also note that this algorithm requires $4n^2$ read/write operations, where n is the number of bits.

```
/*  Total address space is n bits
    For i = 0 to (n – 1); Do
       write 0 in cell (i)
    End
    While i = 0 to (n – 1); Do
       write 1 in cell (i)
       read all cells
          write 0 in cell (i)
    Continue
    End
```

Figure 7.1 GALPAT algorithm to test random access memories (RAMs).

7.2.2 Checker Pattern Test

The checker pattern test writes a pattern of 1 and 0 in alternate cells. After a pause, the whole memory is read. Then, all cells are complemented. Again, after a pause, the whole memory is read. The algorithm is given in Figure 7.2.

This algorithm detects all cell stuck-at-1/0 faults, data retention faults, and 50% of the state transition faults. Address decoder faults and state coupling faults

```
/*  Total address space is n bits
    While i = odd; j = even; Do
        write 0 in cell (i); write 1 in cell (j)
        read all cells
        complement all cells
        read all cells
    Continue
    End
```

Figure 7.2 Checker pattern test to detect RAM faults.

are not covered. The algorithm requires only $4n$ read/write operations. This is a fast algorithm, but fault coverage is not extensive.

7.2.3 Galloping Diagonal/Row/Column Test

This algorithm is based on the principle of GALPAT and is sometimes also called *galloping diagonal*. In this algorithm, instead of shifting a 1 through the memory, a complete diagonal of 1s is shifted. The whole memory is read after each shift. The concept is illustrated in Figure 7.3(a).

10000	01000	00100	00010	00001
01000	00100	00010	00001	10000
00100	00010	00001	10000	01000
00010	00001	10000	01000	00100
00001	10000	01000	00100	00010

Figure 7.3(a) Illustration of the concept of galloping diagonal algorithm to test RAMs.

This algorithm detects all the faults as GALPAT, except for some state coupling faults. However, it is faster than GALPAT and requires $4n^{3/2}$ read/write operations. The variations of this algorithm are *galloping row* and *galloping column*. In galloping row, all the cells in a row contain 1, and this row is shifted through the memory. In the galloping column, all cells contain 1, and this column is shifted through the memory.

In another variation, the whole memory is not read after each shift, but only those cells which are supposed to contain 1. This procedure is known as *shifting diagonal/row/column*. This reduces the number of read/write operations to $4n$.

However, state coupling faults and 50% of the state transition faults are not covered. The procedure for shifting rows when the whole memory is not read is illustrated in Figure 7.3(b).

Note that, if the algorithm given in Figure 7.3b is executed twice, once when background is 0 and second when background is 1, all state transition faults and a majority of state coupling faults will be covered. The number of read/write operations will increase to $8n$, which is still practical.

```
/* Total address space is n bits arranged in k rows
    For i = 0 to (n – 1); Do
        write 0 in cell (i)
    End
    While i = 1 to k; Do
        write 1 in all cells belong to row
        read all cells belong to row
        write 0 in all cells belong to row
    Continue
    End
```

Figure 7.3(b) Shifting Row algorithm to test RAMs.

7.2.4 Marching 1/0 Test Algorithm

The marching 1/0 algorithm is extensively used. In this algorithm, the whole memory is first initialized to either 0 or 1. Then, one cell is read, complemented, and read again. The procedure continues from the first cell to the last cell. When all cells are completed, the whole memory is read. The algorithm is given in Figure 7.4.

This algorithm covers all cell stuck-at-1/0 faults, state transition faults, address decoder faults, and 50% of the state coupling faults. It is significantly fast and requires only $6n$ read/write operations. With a slight modification, as shown in the next subsection, very high fault coverage can be obtained with this algorithm.

7.2.5 Modified Marching 1/0 Test

The marching 1/0 test algorithm can be modified for an extensive fault model. Dekker *et al.* [2–3] have presented $9n$ and $11n$ complexity algorithms to cover 100% of faults. A modified version based on the work by Dekker *et al.* is given in Figure 7.5. This algorithm exploits parallelism and requires $7n$ read/write operations. This algorithm covers all cell stuck-at-1/0 faults, 1-to-0 and 0-to-1 state

```
/* Total address space is n bits
   For i = 0 to (n − 1); Do
      write 0 in cell (i)
   End
   While i = 0 to (n − 1); Do
      read cell (i)
      write 1 in cell (i)
      read cell (i)
      write 0 in cell (i)
      read cell (i)
   Continue
   End
```

Figure 7.4 Marching 1/0 algorithm to test RAMs.

transition faults, bridging faults between two cell (state coupling), data retention faults, and decoder faults.

In this algorithm, the whole memory is first initialized to 0 and the cell value is read. This detects any cell stuck-at-1 fault and also checks state 00 for any two arbitrary cells. During the first cycle, two loops are simultaneously initiated, one addresses from the first cell to the $(n/2)$th cell and other addresses from the last cell to the $(n/2 + 1)$th cell. During this cycle, read output is 1. Hence, any cell stuck-at-0 and transition 0-to-1 faults are detected. During the second cycle, the read output again is 1. This cycle checks state 11 for any two arbitrary cells. Cycle three is similar to cycle one. Two loops are simultaneously initiated, one from the $(n/2 + 1)$th cell to the last cell and the other from the $(n/2)$th cell to the first cell. In this cycle, the read output is 0. Thus, transition 1-to-0 faults are detected. State 01 and 10 for two cells are tested during the first and third cycles. Therefore, any bridging or state coupling faults between two cells are detected. Wait states are used between initialization and the first cycle and between the first and second cycles. These wait states detect data retention faults. Note that a similar kind of wait state can also be used in other algorithms to detect data retention faults.

7.2.6 Comparison and Modification for Word-Oriented Memory

None of the algorithm given above is restricted to a particular architecture. Also, all algorithms are applicable to memory chips as well as memory boards. Any random access memory (SRAM as well as DRAM) can be tested with any algorithm. However, different algorithms are limited to different kinds of faults. A comparison among various algorithms is given in Table 7.1.

```
/* Total address space is n bits
   For i = 0 to (n − 1); Do
      write 0 in cell (i)
      read cell (i)
   End
/* Wait state detects retention of 0. Typical time 100 ms.
   PAUSE
   While i = 0 to (n/2 − 1); j = (n − 1) to (n/2); Do
      write 1 in cell (i)
      read cell (i)
      write 1 in cell (j)
      read cell (j)
   Continue
/* Wait state detects retention of 1. Typical time 100 ms.
   PAUSE
   For i = 0 to (n − 1); Do
      read cell (i)
   End
   While i = (n/2 − 1) to 0; j = (n/2) to (n − 1); Do
      write 0 in cell (i)
      read cell (i)
      write 0 in cell (j)
      read cell (j)
   Continue
   End
```

Figure 7.5 Modified Marching 1/0 test. This algorithm has $7n$ read/write operations.

Table 7.1
Comparison of various RAM test algorithms

Test Method	Complexity	Stuck-at	State Transition	Decoder	State Coupling
GALPAT	$4n^2$	yes	yes	yes	yes
Checker pattern	$4n$	yes	yes	no	no
Galloping Diagonal	$4n^{1/2}$	yes	yes	yes	yes
Marching 1/0	$6n$	yes	yes	yes	no
Modified Marching 1/0	$7n$	yes	yes	yes	yes

All the algorithms in the form given above are applicable to bit-oriented memory. However, these can be modified to m-bit-wide word-oriented memory. By using K different data backgrounds, where $K = \log_2 m$, any algorithm can be used for word-oriented memory. For example, for byte-oriented memory, we need three different data backgrounds, i.e.:

First data background—10101010;

Second data background—11001100;

Third data background—11110000.

Also note that a bit-oriented memory can be designed such that it appears to be byte-oriented during the test mode. Thus, using different data backgrounds, total test time can be reduced by a factor of 8/3. In general, if a bit-oriented memory is tested as an m-bit-wide word-oriented memory, test time is reduced by a factor of $m/\log_2 m$, which is a significant savings.

7.3 TESTABLE DESIGNS

In the preceding section, various test algorithms are given for RAMs. As the memory size increases to multimegabits, built-in testing becomes a necessity. Even an $O(n)$ algorithm requires an excessive amount of time to be feasible. Assuming that a tester is running at 10 MHz, a 16-Mbit memory will require about 11.2 seconds if we use $7n$ algorithm, but about 3.25 years if we use GALPAT. The estimated test time for various algorithms are given in Table 7.2.

Table 7.2
Estimated test time for different sizes of memories using various algorithms.

Test Method	Complexity	1 Mbits	16 Mbits	64 Mbits
GALPAT	$4n^2$	111 hr	1185 days	52 yr
Checker Pattern	$4n$	.4 s	6.4 s	25.6 s
Galloping Diagonal	$4n^{3/2}$	400 s	7.1 hr	56.9 hr
Marching 1/0	$6n$	.6 s	9.6 s	38.4 s
Modified Marching 1/0	$7n$	.7 s	11.2 s	44.8

Testable designs have been developed to reduce the test time. The basic philosophy behind these testable architectures is to partition a memory into small blocks and to test them in parallel.

7.3.1 BIST Memory

Section 7.2.6 mentions that a test algorithm can be used for word-oriented memory using different data backgrounds. In a word-oriented memory (word size-m bits), m bits are tested in parallel. Therefore, if a bit-oriented memory is designed such that it appears as a word-oriented memory during the test phase, test time can be reduced by a factor of $m/\log_2 m$. However, note that the implementation of such procedure at the chip level requires a modification in architecture. In general, memory address generation logic and a data bus are designed with additional control logic. A generalized BIST memory architecture is given in Figure 7.6 [9].

You and Hayes [10] suggested a BIST design using on-chip test generation and response evaluation. In this design, the whole memory is divided into multiple blocks, while all memory cells in a block form a circular shift register during test mode. The on-chip logic scans all the blocks in parallel by concurrently shifting the same data in different blocks. The scanned data or responses from different

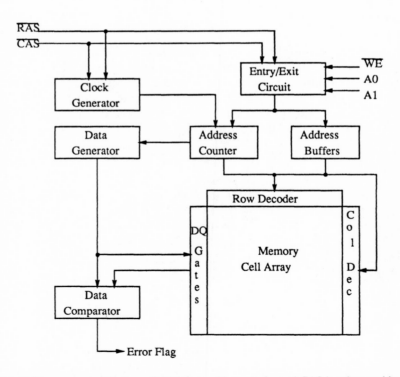

Figure 7.6 BIST memory architecture: RAS is row address strobe and CAS is column address strobe. (From [9], © 1987 IEEE. Reprinted with permission.)

blocks are compared using on-chip logic to detect a fault. The memory architecture for two blocks is shown in Figure 7.7. This architecture eliminates the need of external test generation. However, the hardware overhead in this design is high. It has been reported that for 1-Mbit DRAMs, the hardware overhead is about 5%. Another disadvantage is due to the modification in sense amplifiers and refresh logic, whereby this design causes a significant performance penalty.

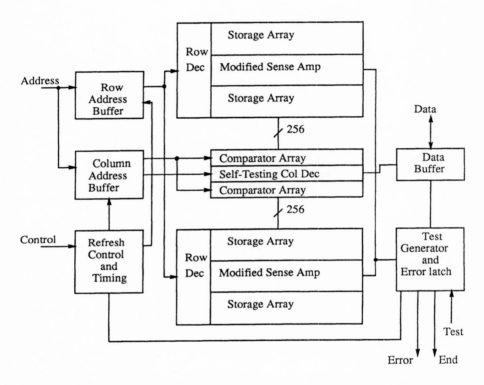

Figure 7.7 BIST DRAM architecture. (From [10], © 1985 IEEE. Reprinted with permission.)

7.3.2 Memory Partitioning Methods

Jarwala and Pradhan [7] have suggested partitioning the memory into small blocks and accessing them in parallel using a data bus. The memory is partitioned into small blocks making an H-tree using test comparators and switches between the blocks. The design is shown in Figure 7.8. In this architecture, a memory can be partitioned into as many blocks as desired. However, the hardware overhead is

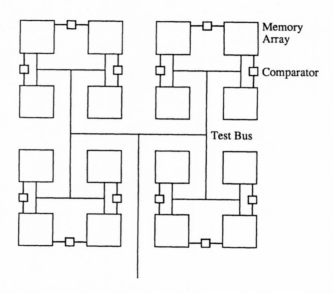

Figure 7.8 TRAM architecture when memory is partitioned into 16 blocks. (From [7], © 1988 IEEE. Reprinted with permission.)

significant due to the use of the data bus for test application. For 1-Mbit RAMs, the hardware overhead becomes 15% when memory is partitioned into eight blocks, each 128-kbits.

Sridhar [6] has suggested partitioning a memory into small blocks and use an on-chip kbit-wide *parallel signature analyzer* (PSA) to access kbits simultaneously. The overall design is shown in Figure 7.9(a), and the PSA design is given in Figure 7.9(b). The PSA is designed to operate in three different modes:

1. *The scan mode*: During the scan mode, the PSA works as a simple shift register. It is loaded serially with a seed pattern to generate test. At the end of test, the contents of PSA (kbit signature) is scanned.
2. *The write mode*: During the write mode, the contents of the PSA is written to a number of memory cells (kbit) in parallel.
3. *The signature/read mode*: During this mode, the contents of memory cells (kbit) are read and a new kbit signature is generated. This signature is used to determine whether a memory has a fault.

The hardware overhead is reported about 2.2% to 3.1% in a 64-kbit SRAM using 64-bit PSA. Apart from the problem of aliasing in PSA, the hardware overhead becomes significant in a multimegabit RAM.

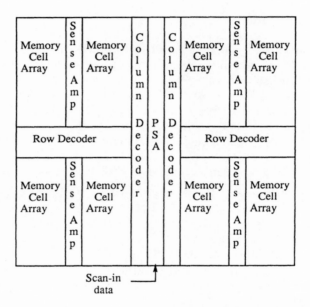

Figure 7.9(a) Testable memory architecture using PSA. (From [6], © 1986 IEEE. Reprinted with permission.)

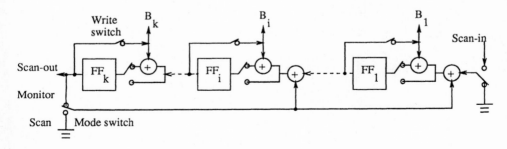

Figure 7.9(b) Block schematic of PSA used in Figure 7.9(a). (From [6], © 1986 IEEE. Reprinted with permission.)

7.3.3 STD Architecture

In the STD architecture, the main idea is to partition the memory address decoder into multiple-levels and design the memory accordingly [8]. Thus, the whole memory is partitioned by itself without large hardware overhead. The only requirement is a modified decoder for the most significant address lines.

Below, four examples are given to illustrate the STD architecture. These examples cover both 1-bit and m-bit word-size memories, and allow memory size to increase by a factor of 2^x. To express the memory size, the notation $n \times m$ is used, where n represents the number of words and m represents the word size. For example, 16k × 1 represents 16-kbits and 16k × 8 represents 16-kbytes.

Example 1: 8k × 1 Memory. This memory requires a 13-to-8k decoder, which can be implemented at two levels, using 10-to-1k decoder and a 3-to-8 decoder. Thus, the memory can be designed by eight blocks of 1-kbits (each associated with 10-to-1k decoder) and a 3-to-8 decoder.

The design is made testable by modifying the 3-to-8 decoder, which contains the most significant address lines A10–A12. The modification is done by adding one extra control signal to the decoder. To illustrate the concept, the design of 2-to-4 decoder is given in Figure 7.10. Also, a parity circuit is added at the outputs of the 1-kbit blocks. The testable design is given in Figure 7.11. The control signal added to the 3-to-8 decoder makes possible the selection of all the decoder output lines when control signal $c = 1$ (this is done during the test mode). When $c = 0$, the decoder is in its normal mode and selects only one of its outputs.

To test this memory, the control signal c is kept to 1. Thus, the same data read/write operations can be done to all eight memory blocks using address lines A0–A9. During this mode, all eight blocks are tested in parallel. Note that in the

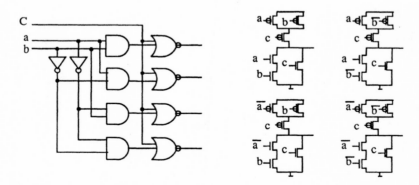

Figure 7.10 Circuit diagram of modified 2-to-4 decoder with an external control c.

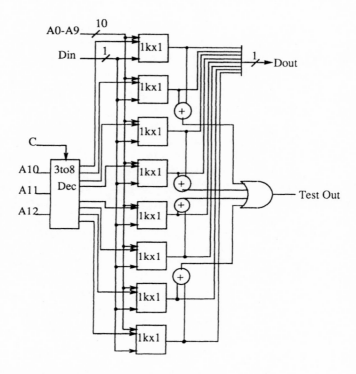

Figure 7.11 Testable design of 8k × 1 memory (Example 1).

case of a fault in any block, the output of the parity circuit would be 1, and hence the fault is detected. By using the modified marching 1/0 algorithm, all eight blocks can be tested by 7k read/write operations.

After the testing of memory blocks, the control signal c is switched to 0, converting the 3-to-8 decoder into normal mode. Under this situation, eight input combinations are needed to test the 3-to-8 decoder. Note that if 1k × 1 blocks are tested by the modified marching 1/0 algorithm, all cells contain 0 after the conclusion of the test. Hence, for each combination of address lines A10–A12, we can detect decoder faults by writing a 1 in a cell and reading it. However, to test the next combination, the cell should contain 0. Thus, 24 read/write operations are required (the last write is not necessary; thus, 23 operations are enough). These eight possible combinations are applied at the address lines A10–A12, while keeping A0–A9 at a fixed value (preferably, all 0s or all 1s). The response is observed at D_{out} line. Thus, the whole 8-kbit memory can be tested in the test time necessary to test 1-kbit memory, (7k + 24) read/write operations to be accurate. The hardware overhead in this design is one control signal, 4 XOR gates, and 1 OR gate. Because the control line is limited to the 3-to-8 decoder and does not extend to the memory blocks, the routing area is negligible for the control signal. The 10-to-1k decoders within the memory blocks can also be implemented in two-level or multilevel. However, these decoders need no modification because partition size is 1-kbit.

Example 2: 32-kbit Memory. This memory can be built by four blocks of 8-kbit memories, each equivalent to Example 1, and an additional 2-to-4 decoder. This 2-to-4 decoder contains the most significant address lines, and hence it is modified by a control signal. The control signal used in the 3-to-8 decoder in Figure 7.10 can be used in the 2-to-4 decoder, as shown in Figure 7.12. During the test mode, the four 8k × 1 blocks are selected using the control signal ($c = 1$) and all eight 1k × 1 blocks are selected for each 8k × 1 block. Therefore, by setting $c = 1$, all 32 blocks of 1k × 1 memory are selected. These 32 blocks are tested in parallel by 7k read/write operations using address lines A0–A9.

After testing of the memory blocks, the control signal is switched to 0. Under this condition, the 2-to-4 decoder and four 3-to-8 decoders are tested by 32 combinations (96 read/write operations). Hence, whole 32-kbits memory is tested as 1-kbits memory using (7k + 96) vectors. If the control signals to 2-to-4 decoder and 3-to-8 decoders are different, then all four 3-to-8 decoders can also be tested in parallel. In that case, total number of test vectors is (7k + 36).

Extra hardware required in this 32k × 1 design is four parity circuits, each having four XOR gates and one OR gate, another OR gate and a control signal. The routing area of the control signal is again negligible. Effectively, hardware overhead in this case is sixteen XOR gates and five OR gates.

The above two examples are restricted to bit-oriented memory where the data line is only one bit wide and observability is poor. When the word size is *m*-

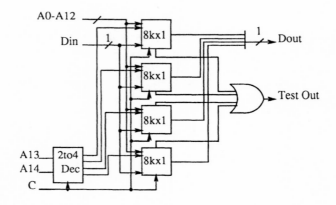

Figure 7.12 Testable design of 32k × 1 memory.

bits, *m* data lines are observable. Therefore, we can easily divide the whole memory into *m* blocks and test them in parallel.

Example 3: 8k × 8 (8-kbyte) Memory. We can design an 8k × 8 memory by using eight blocks of 8k × 1 memory (Figure 7.13). In this design, each block of 8k × 1 memory is equivalent to the circuit given in Figure 7.11.

All eight blocks of 8k × 1 memory can be tested in parallel by making $D_{in0} = D_{in1} = \ldots = D_{in7}$. Effectively, this testing procedure is the same as in Example 1. The number of memory operations is (7k + 24). Because each test output is separately observable, any fault can be detected and faulty blocks can be uniquely identified. After testing of the memory blocks, eight additional write and read operations (sixteen operations) are used to test the coupling among D_{in} lines or among D_{out} lines. This is achieved by keeping one D_{in} line at 1 while all others are at 0, for a fixed value of A0–A12. Thus, the whole 8k × 8 memory is tested by (7k + 40) vectors.

The hardware overhead in this case is eight parity trees, each having four XOR gates and one OR gate (total 32 XOR gates and 8 OR gates), and one control

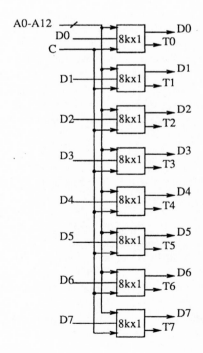

Figure 7.13 Testable design of 8k × 8 memory.

signal. Note that modification of the test algorithm to a byte-oriented memory, as given in Section 7.2.6, is not desirable. The amount of hardware overhead will remain the same, but the number of test vectors will increase to (21k + 24).

The final example illustrates the case when the number of words is increased for a memory of *m*-bit word size.

Example 4: 32k × 8 Memory. This memory can be built by using four blocks of 8k × 8 memory and a 2-to-4 decoder. The concept is similar to that of Example 2. In this case, each block of 8k × 8 memory is equivalent to Figure 7.12. The additional 2-to-4 decoder, which contains the most significant address lines, is modified by adding a control signal. If we consider a separate control signal than used in 8k × 8 blocks, the memory can be tested by (7k + 48 + 12) vectors or read/write operations. The hardware overhead in this case is two control signals and [4 × (32 XOR + 8 OR) + 8 OR = 128 XOR + 40 OR] gates.

Another possibility is to design this memory by using eight blocks of 32k × 1 memory. The concept is similar to that of Example 3. However, in this case, each block of 32k × 1 memory is equivalent to Figure 7.13. With this design, a 32-kbyte memory can be tested by (7k + 96 + 12) vectors. The hardware overhead is one control signal and [8 × (16 XOR + 5 OR) = 128 XOR + 40 OR] gates.

In both approaches, the hardware overhead and test time are comparable. The amount of hardware overhead can be reduced significantly by using bigger blocks of memory arrays instead of 1k × 1 arrays. However with the larger blocks, memory test time will increase. For example, if 8k × 1 memory blocks are used, the whole memory can be tested by (56k + 12) vectors, while only (16 XOR + 8 OR) gates are required (see Figure 7.14). Note that, equivalently, 256k × 1 memory can be designed by four additional 3-to-8 decoders, and can be tested by (56k + 48) read/write operations.

7.4 FAULT DIAGNOSIS AND RECONFIGURATION

With slight modifications in the designs given in Section 7.3, memory can be designed for fault diagnosis. The basic idea is to use a register instead of a parity circuit to obtain better observability. Consider the design of 8k × 1 memory as given in Example 1. The modified design is shown in Figure 7.15. The design given in Figure 7.15 is basically the same as given in Figure 7.11, except that the OR gate in parity circuit of Figure 7.11 has been replaced by a 4-bit register. The test method is the same as given in Section 7.3.3. However, instead of looking for one-bit parity output, the test responses of two blocks (each 1k × 1) are observable. Observability can be further enhanced if the whole parity circuit is replaced by an 8-bit register. In this case, a faulty 1k × 1 block can be uniquely identified. In

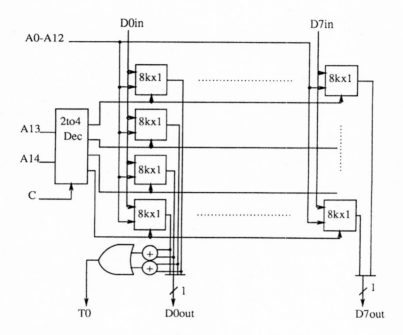

Figure 7.14 Testable design of 32k × 8 memory.

general, if only fault detection is required, parity circuits are recommended because the area required by the registers is greater than for XOR and OR gates.

After locating a faulty block, the memory can be reconfigured into 3/4, 1/2, or 1/4 of the original capacity. The idea is permanently connecting the corresponding address line of the most significant part of the address decoder to the GND or the V_{dd}. For example, two blocks (each 1k × 1) can be disconnected in Figure 7.15 by connecting the corresponding address line (A10–A12) to the GND/V_{dd}. Also possible is to reconfigure the memory into 1/2 of the original capacity.

7.5 ADVANTAGES AND DISADVANTAGES

The main advantage of partitioning methods is that very small test time can be achieved for any size of memory. Another advantage is that any fault model can be assumed and an appropriate test algorithm can be used. In general, the partitioning methods are highly structured and test vectors need not be calculated for different memories having the same size of partitions (except for a few additional

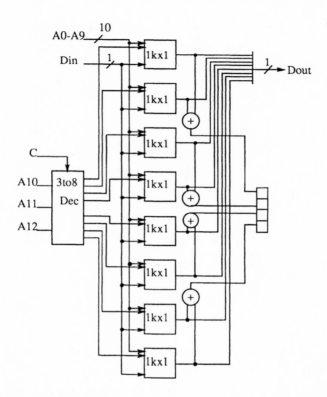

Figure 7.15 An 8k × 1 memory for fault diagnosis and reconfiguration.

vectors). With these architectures, large memories can be designed in significantly less time using existing memory blocks of small sizes. Note that these architectures are not limited to the chip design. A memory board can be designed without any modification in the architecture.

As discussed in Section 7.4, fault diagnosis can also be achieved during testing. Fault diagnosis is extremely important from a reliability point of view. In the case of a memory board, this is highly desirable. By identifying a faulty memory block (memory chip in the case of a board), it can be replaced by a good block if redundancy is available. If redundancy is not available, the faulty block can be disconnected and the most significant decoder is modified so that good memory blocks can still be used. The memory subsequently can be configured into 3/4, 1/2, or 1/4 capacity of the original size.

The negative aspect of the BIST architecture or partitioning methods is the requirement of extra hardware. In the BIST architecture, hardware overhead is significant, making these designs impractical. In partitioning methods for large memories, if the test time is kept extremely small (for example, a 4-Mbit memory partitioned into 1-kbit blocks), the hardware overhead becomes significant, and it is inversely proportional to the partition size. The hardware overhead is minimal in the STD architecture. The trade-off for various memories is shown in Figure 7.16 for the STD architecture. Note that, in Figure 7.16, the overhead is measured in terms of the number of gates. Generally, overhead is expressed in terms of percentage, i.e., as a portion of memory size. However, the percentage of overhead is an inadequate representation because it understates the actual situation. For example, 10% overhead for 16-kbit memory might be acceptable, but for 16-Mbit memory it is highly undesirable. The partitioning and implementation of a decoder into multilevel in STD architecture results in a reduction in transistor count. This fact can be visualized by considering a small example, a 4-to-16 decoder. One level implementation of a 4-to-16 decoder using 4-input gates requires 128 transistors (64 n-MOS and 64 p-MOS transistors). The same decoder can be implemented at two levels using five 2-to-4 decoders. In this case, the total number of transistors is $(5 \times 16 = 80)$.

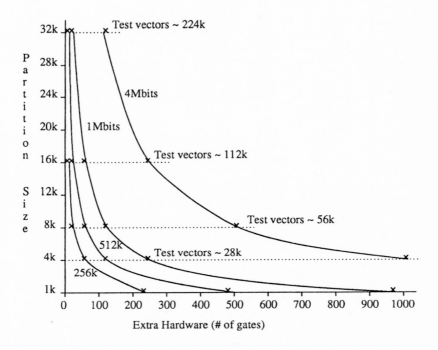

Figure 7.16 Trade-off in partition size versus extra hardware. Dotted lines represent constant test time.

Another disadvantage in BIST designs and partitioning methods is the penalty in performance. The extra hardware increases parasitic line capacitance, which causes a larger delay during normal operation. In some designs, sense amplifiers and address decoders are modified. This also increases the access time during normal operation. The performance penalty is minimal in the STD architecture. Partitioning of the decoder in multilevel results in a decrease in signal propagation delay. Therefore, such partitioning is desirable for improving the performance.

7.6 SUMMARY

The testing of random access memory is extremely time consuming. For very large memories, test time becomes the critical factor in overall production cost. The problem of excessively large test time cannot be solved alone by a clever test algorithm.

This chapter gives the most widely used test algorithms for RAMs. Different variations are discussed and various algorithms are compared. In general, a user is advised to choose the $O(n)$ algorithm, unless use of a higher complexity algorithm is specified. The another selection criterion is 100% fault coverage under a specified fault model.

The various design methods are discussed to help in RAM testability. The basic idea behind these testable architectures is to partition the whole memory into small blocks and to test them in parallel. Each implementation has its own advantages and disadvantages. Based upon hardware overhead, we recommend use of the STD architecture. Finally, we discussed fault diagnosis and memory reconfiguration method.

PROBLEMS

1. A 64-bit RAM is implemented in an 8×8 array. Compute the complete test sequence using (i) checker pattern test, (ii) shifting column test, and (iii) marching 1/0 test.
2. Give the complete test sequence for 64-byte RAM using modified marching 1/0 test algorithm.
3. Suppose that you have access to $1k \times 1$ RAM chips. Design a 8-kbit testable RAM module using (i) You and Hayes BIST design, and (ii) Sridhar's partitioning architecture.
4. Design a $64k \times 16$ RAM chip using STD architecture when the partition size is 4 kbytes. Redesign the same capacity bit oriented memory of 1 Mbit using partition size of 16 kbits.

REFERENCES

1. R. Nair, S.M. Thatte, and J.A. Abraham, "Efficient Algorithms for Testing Semiconductor Random Access Memories," *IEEE Trans. Comp.*, **27**(6), pp. 572–576, June 1978.
2. R. Dekker, F. Beenker, and L. Thijssen, "A Realistic Fault Model and Test Algorithm for Static Random Access Memories," *IEEE Trans. CAD*, **9**(6), pp. 567–572, June 1990.
3. R. Dekker, F. Beenker, and L. Thijssen, "Fault Modeling and Test Algorithm Development for Static Random Access Memories," *Proc. Int. Test Conf.*, pp. 343–352, 1988.
4. J.P. Hayes, "Detection of Pattern Sensitive Faults in Random Access Memories," *IEEE Trans. Comp.* **24**(2), pp. 150–157, Feb. 1975.
5. J.P. Hayes, "Testing Memories for Single Cell Pattern Sensitive Faults," *IEEE Trans. Comp.*, **29**(3), pp. 249–254, March 1980.
6. T. Sridhar, "A New Parallel Test Approach for Large Memories," *IEEE Design and Test*, pp. 15–22, Aug. 1986.
7. N.T. Jarwala and D.K. Pradham, "TRAM: A Design Methodology for High Performance, Easily Testable, Multimegabit RAMs," *IEEE Trans. Comp.*, **37**(10), pp. 1235–1250, Oct. 1988.
8. R. Rajsuman, "An Apparatus and Method for Testing Random Access Memories," US Patent Application, Serial Number 620359, Nov. 20, 1990.
9. T. Ohsawa *et al.*, "A 60-ns 4-Mbit CMOS DRAM with Built-in Self-Test Function," *IEEE J. Solid State Circuits*, **22**(5), pp. 663–668, Oct. 1987.
10. Y. You and J.P. Hayes, "A Self Test Dynamic RAM Chip," *IEEE J. Solid State Circuits*, **20**(1), pp. 428–435, Feb. 1985.

Chapter 8
Testing of Sequential Circuits

8.1 INTRODUCTION

Fault detection in sequential circuits implementing a *finite state machine* (FSM) is much more complex than fault detection in a combinational circuit. These circuits contain memory elements and some feedback lines. In both the Moore and Mealy models, the output of such circuits is a function of the present state and the input. In general, the feedback lines in the circuit are not directly observable as well as hard to control. Because of the feedback lines, the output of the circuit cannot be specified by only the inputs. To determine the output, one needs to know the present state of the memory elements and the logic values at the feedback lines. Thus, a fault in a sequential circuit is not detected within one time frame by simply applying an input vector and observing the output.

Several test generation methods have been proposed in the literature for sequential circuits. One possible solution is to design a circuit for easy testing. Scan design is a popular method used to achieve testability by using extra hardware. We discuss methods to provide testability through extra hardware in Chapter 10. In this chapter, we discuss the methods to test sequential circuits that are not designed with extra hardware while considering testability features in the design.

8.2 TESTING PROBLEM IN SEQUENTIAL CIRCUITS

The test generation for sequential circuits becomes complex because of the following requirements:

1. To set the memory elements into a known state before the actual test application or fault sensitization;
2. To select an input vector and present state such that a fault can be sensitized;

---

3. To select inputs such that the fault-effect propagates toward the primary outputs, either directly or through the feedback loops.

When a sequential machine is powered on, it assumes an unknown state. Thus, unless it is initialized into a known state, it cannot be tested. After initialization, due to the presence of memory elements and feedback loops, more than one vector may be required to sensitize a fault. Similarly, a sequence of vectors may be needed to propagate the fault effect to an observable output.

This problem is further complicated by the requirement that a signal transition must occur within a fixed duration relative to other signals. If a signal transition does not take place within the correct interval, a flip-flop may spuriously set or reset. Although all components may be fault-free, the circuit may still malfunction in this situation. Also, some flip-flops have invalid input combinations; great care must be taken during testing that such combinations do not occur. For example, a cross-coupled NOR gate latch should not have an input of 11. A timing problem may cause such inputs and set the circuit into an unspecified state. Also, some input combination may cause the circuit to oscillate. A discussion of timing faults, races, and hazards is beyond the scope of this text. Here, we discuss the testing of logical faults in sequential circuits which are free of timing errors. Also, asynchronous sequential circuits are not explicitly considered. However, the concepts discussed here can be extended to them.

8.3 STATE TABLE APPROACH

State table and *state transition graph* (STG) are popular representations of a finite state sequential circuit. An obvious testing method is to verify the state table or state transition graph. In this procedure, the circuit is first initialized into a known state (reset state) and then all state transitions (edges in the STG) are verified. One difficulty of this method is initializing the circuit. A possible solution is to design the circuit with an extra control (set/clear signal), which can initialize all the flip-flops. The use of an extra control is a simple as well as effective method, although it creates penalty in hardware. However, in the present technology, all flip-flops (including standard cells) are designed with a set/clear signal. A sequential circuit can thus be initialized easily to all 0 states.

If a circuit does not have a set/clear control, it can only be initialized by an input sequence that will force it into a certain state. In the next section, we discuss methods to find initializing sequences.

8.3.1 Initialization of Sequential Circuits

A difficult but important task in the testing of sequential circuit is to initialize it into a known state. The objective is to apply a sequence of input vectors such that the circuit's final state can be uniquely determined, regardless of its initial state. Before we discuss how it is done, some terms need to be defined.

Initial Uncertainty. An initial uncertainty is the set of states that a circuit can take when it is powered on. In the worst case, if a circuit has n possible states, it can be in any one of these n states.

Strongly Connected Circuit. A sequential circuit is called strongly connected if, for every pair of states S_i and S_j, an input sequence is available that takes the circuit from state S_i to state S_j.

8.3.1.1 Synchronizing Sequence

A synchronizing sequence is an input sequence that forces a fault-free sequential circuit to a specific state regardless of its initial state. It is an ideal initialization sequence. However, in most cases without a specific design, it is difficult to obtain a synchronizing sequence for an arbitrary sequential circuit. Note that if all flip-flops have a set/clear signal, we can consider the presence of a synchronizing sequence of length 1.

To obtain a synchronizing sequence, a successor tree or *binary decision diagram* (BDD) is made. Each node in the BDD is associated with an uncertainty vector, while the first node contains an initial uncertainty. The number of children at each node is equal to the number of possible input combinations to the circuit. A node is considered to be a terminating node with success if its uncertainty vector contains only a single element. If a child node has the same uncertainty vector as its preceding node, it is considered to be a terminating node without success. A synchronizing sequence provides a path from initial uncertainty to the terminating node with a single state.

Example 1. Consider the sequential circuit M1, with state table as given in Table 8.1. The state diagram for this circuit is given in Figure 8.1. This circuit has four possible states, one primary input, and one primary output. The successor tree for

Table 8.1
State table of a sequential circuit M1.

Present State	Next State and Output	
	Input 0	Input 1
A	C,1	A,0
B	D,0	C,0
C	B,1	A,0
D	A,0	B,1

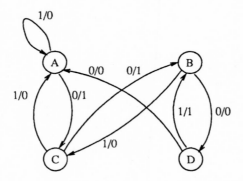

Figure 8.1 State diagram of the circuit used in Example 1.

this circuit to obtain a synchronizing sequence is shown in Figure 8.2. A three-vector input sequence 111 forces this circuit to state A, regardless of its present state. Thus, input 111 is a synchronizing sequence. As this procedure is adaptive, we can terminate the process after the second input if the output response under the first two vectors is 00. In the worst case, we need to apply all three vectors.

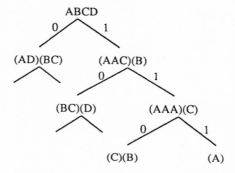

Figure 8.2 Synchronizing tree for the circuit given in Figure 8.1.

8.3.1.2 Homing Sequence

A homing sequence is an input sequence under which the response of a fault-free sequential machine is sufficient for knowing its final state, regardless of its initial state [1]. The homing sequence is also obtained through a successor tree. In a homing sequence successor tree, a node is considered to be a terminating node without success if a child node has the same uncertainty vector as the preceding node. If a node has an uncertainty vector with unique states, it is considered to be a terminating node with success. A homing sequence provides a path from initial uncertainty to a terminating node with uniquely differentiated states. Although computationally expensive, all reduced sequential circuits have at least one homing sequence [2]. As shown in [2], for any reduced n-state sequential circuits ($n \leq 2^k$, where k is the number of flip-flops), the upper bound on the length of homing sequence is $\leq (1/2)[n(n - 1)]$.

Example 2. The sequential circuit M1, as described in Table 8.1, has a homing sequence 011 as shown in Figure 8.3. The response of the circuit is given in Table 8.2. The output produced in response to 011 is sufficient to determine its final state. Application of input sequence 011 forces this circuit to either state A or C, which are differentiated by the output. It should be noted that we cannot determine the initial state because of an ambiguity between initial states A and C, as the outputs are the same. Again this process is adaptive, and we can terminate it as soon as we eliminate the uncertainty. For example, if the circuit originally is in either state B or D, the response to the first two vectors {01} is sufficient to determine its final state as either A or B.

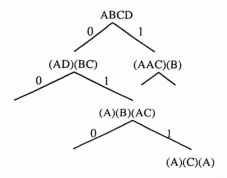

Figure 8.3 Homing tree for the circuit shown in Figure 8.1.

8.3.1.3 Distinguishing Sequence

A distinguishing sequence is an input sequence under which the response of a fault-free sequential circuit is sufficient for knowing its final state as well as initial state,

Table 8.2
Response of circuit M1 to input sequence 011.

Present State	Output	Final State
A	1 0 0	A
B	0 1 0	C
C	1 0 0	A
D	0 0 0	A

regardless of its initial state [1]. In other words, the circuit output in response to a distinguishing sequence is different for each initial state. Note that every distinguishing sequence is also a homing sequence, but the converse is not true. The procedure to obtain a distinguishing sequence is the same as that of a homing sequence. However, a terminating node with success in distinguishing sequence BDD is a node with uncertainty vector having only a single component.

Example 3. Consider the circuit M1 under input sequence {00}. The BDD for this case is shown in Figure 8.4 and the response is given in Table 8.3.

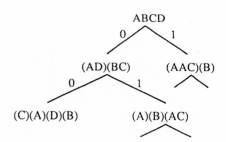

Figure 8.4 Distinguishing tree for the circuit of Figure 8.1.

The output in response to 00 is sufficient to determine the circuit's final state. For each initial state, the output is unique, and hence the circuit's initial state can also be determined. Thus, {00} is a distinguishing sequence.

In general, the computational complexity of a distinguishing sequence is higher than that of a homing sequence. Note that many sequential circuits may not have a distinguishing sequence. Any circuit M, which does not have a distinguishing sequence, can be implemented as circuit M'[1] such that M' contains M, and M' has a distinguishing sequence of length $\leq (1/2)[n(n-1)]$. The basic

Table 8.3
Response of circuit M1 to input sequence 00.

Present State	Output	Final State
A	1 1	B
B	0 0	A
C	1 0	D
D	0 1	C

procedure to obtain M′ is to add some extra logic to M, making observable the outputs of some flip-flops.

8.3.2 State Table Verification

A possible method for testing a sequential circuit is to check whether the circuit is operating according to its state transition table or state transition graph. This is equivalent to functional testing and in most of the cases works fairly well. In this approach, the circuit should have at least one synchronizing or homing sequence to initialize the circuit. As stated above, a reduced, strongly connected circuit always has a homing sequence. Thus, an input sequence can be obtained to verify all state transitions in a reduced, strongly connected machine.

Using a homing sequence, the circuit is first initialized. While the circuit is in a known state (initialized state), an input sequence is applied such that the machine will visit all the states while covering all specified state transitions. The output of the circuit is observed very closely. Any deviation at any instant from the expected output shows a fault in the circuit. Although this procedure is not guaranteed, experience with sequential circuits dictates that in a majority of cases it is sufficient. For example, for the circuit given in Figure 8.1, after synchronizing to state A, a sequence of 11 vectors checks the functionality of the circuit completely. One possible test sequence is {1 0 0 0 1 1 1 0 0 0 0}. It should be noted that testing based upon a state table or state diagram detects only structural faults that cause functional errors. The circuit may have some line stuck-at faults and transistor-level faults, which cannot be detected by such testing.

If a circuit has a distinguishing sequence, a more elaborate testing procedure can be developed. In the literature, such methods are known as *checking experiment* [3–4]. First, the circuit is initialized by applying a homing or distinguishing sequence. To check a particular transition from state S_i to S_j, a transfer sequence is applied, which brings the circuit to state S_i from the initialized state. While the circuit is in state S_i, an appropriate input is applied such that a fault-free circuit will go to state S_j. Finally, a distinguishing sequence is applied to verify that the circuit indeed goes to state S_j. This procedure is executed for every pair of states S_i and S_j to verify all the transitions individually. The detailed procedure is given as follows [1]:

1. Apply a *homing sequence* (HS) to initialize the circuit into state S_i.
2. Apply a *transfer sequence* (TS) to bring the circuit into initial state S_0.
3. Apply the shortest *distinguishing sequence* (DS) to verify state S_0.
4. If the circuit is not in S_0 after step 3, apply a TS to obtain S_0.
5. As long as a new transition is covered, apply an input 1.
6. If an input 1 does not cover any new transition, apply a 0, and then repeat step 5.
7. When steps 5 and 6 do not cover any new transition and all transitions are not covered, apply a TS from S_i to S_k, where some transition of S_k is not covered.
8. Repeat steps 5 to 7 until all transitions are covered.

Note that this procedure not only detects a fault, but it also identifies the corrupted edge in the state diagram. Using this procedure, any n-state sequential circuit, which has a distinguishing sequence of length L, can be tested by a sequence of length $\leq [nm + n(m - 1)L + L + (m - 1)(n - 1)^2]$, where m is the number of inputs. If $L \leq (1/2)[n(n - 1)]$, the length of a checking experiment is bounded by $\leq (m - 1)(n - 1)^3$.

The major advantage of such testing is that a detailed circuit description is not required. A test engineer does not need a gate-level description and can work only with the functionality of the circuit in terms of a state table or state transition graph. This is effectively a "black box" approach, verifying the input-output relationship.

One disadvantage is that the methods based upon the state table are suitable only for small circuits. For large circuits, the memory requirement to store large tables makes these methods impractical. Another problem with these methods is the length of the test sequence. Even for moderately sized circuits, the length of a test sequence in the checking experiment becomes so large as to be impractical.

8.4 GATE LEVEL TEST GENERATION METHODS

The test methods based upon verification of state table or state transition graph detect functional error when an edge in STG is corrupted. Many faults that do not change a transition remain undetected by such testing. These faults create severe reliability problems. Better fault coverage can be obtained by testing for all line stuck-at faults. A detailed gate-level circuit description is necessary for such testing. Test generation methods for stuck-at faults for pure combinational circuits are given in Chapter 5. In this section, we will extend these methods to detect stuck-at faults in sequential circuits.

8.4.1 Sequential Test Generation by Boolean Difference

In Section 5.2.1, we discussed the Boolean difference method to generate test vectors for pure combinational circuits. To extend this method for sequential circuits, we define the Boolean difference of the output Z_j with respect to line x_i at time k as follows [5]:

$$\frac{dZ_j}{dx_i} = Z_j(X_{i1}, S^k) \text{ XOR } Z_j(X_{i0}, S^k) \tag{8.1}$$

where

$$X_{i1} = (x_1, \ldots, x_i, \ldots, x_n)$$

and

$$X_{i0} = (x_1, \ldots, \overline{x_i}, \ldots, x_n)$$

Where S^k is the state of circuit at time k, and $x_1, \ldots, x_n$ are the inputs. If S^k is not present, Equation (8.1) represents the Boolean difference of a pure combinational circuit. The right-hand side of Equation (8.1) is written as $[Q(X_i, S^k)]$, where Q denotes a function of X_i and S^k, and X_i represents an input vector, which does not contain x_i.

If $G_1, G_2, \ldots, G_b$ are the gates in the path from x_i to Z_j, we represent this path P_i as follows:

$$(P_i) = (x_i, G_1, \ldots, G_b, Z_j)$$

1. For a nonredundant nonreconvergent path P_i of a sequential circuit, a complete test set is given by dZ_j/dx_i.
2. For a nonredundant reconvergence path P_i, test set is given by Boolean difference chain $\nabla(P_i)$ if the circuit can be placed in state S^k. The Boolean difference chain $\nabla(P_i)$ is given as

$$\nabla(P_i) = \frac{dG_b}{dG_{b-1}} \frac{dG_{b-1}}{dG_{b-2}} \cdots \frac{dG_2}{dG_1} \frac{dG_1}{dx_i} \tag{8.2}$$

Based upon the above two points, a test sequence can be generated as follows:

1. For each output Z_j, compute Boolean difference chain $\nabla(P_i)$ from input $x_i (i = 1, 2, \ldots, n)$ to Z_j.
2. If $\nabla(P_i)$ contains more than one term, select the term for which $\nabla(P_i) \cdot x_i \neq 0$ and $\nabla(P_i) \cdot \overline{x_i} \neq 0$.

3. If the term from step 2 contains one or more feedback variables that form a state, calculate the homing sequence that should be applied before the test pattern.
4. If a state does not exist, test patterns can be used without HS. The test patterns are selected terms from $\nabla(P_i)$ with x_i and $\overline{x}_i$.

This procedure is illustrated with an example.

Example 4. Consider the circuit shown in Figure 8.5. The output function for this circuit is given by $F = ab + af + \overline{b}f$, where f is the feedback variable. Consider the path from a to F. Since this path is reconvergent, we use a Boolean difference chain:

$$\nabla(P_i) = \nabla[a, G_1, G_2, F]$$

or

$$\nabla(P_i) = (b)(\overline{a}b + \overline{f}) = \overline{a}b + b\overline{f}$$

Here, we have two possible terms, i.e., $(\overline{a}b)$ and $(b\overline{f})$. As $(\overline{a}b) \cdot da = 0$, we select $(b\overline{f})$. This term has a feedback variable; hence, we calculate an HS as $\overline{a}b$. Thus, the test sequences are:

1. $\overline{a}b$ (HS), $(b\overline{f}) \cdot a$; or $ab = [01, 01]$
2. $\overline{a}b$ (HS), $(b\overline{f}) \cdot \overline{a}$; or $ab = [01, 11]$

For path $P_2 = (b, G_1, G_2, F)$, we have a Boolean difference chain $\nabla(P_2) = (a)(\overline{a}b + \overline{f}) = a\overline{f}$. Thus, the test sequences are $ab = [01, 11]$ and $[01, 00, 10]$. Note that, in the last sequence, an additional pattern 00 appeared because the HS length is 2.

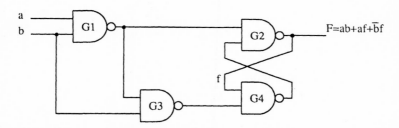

Figure 8.5 A sequential circuit to illustrate test generation by Boolean difference. (From [5], © 1971 IEEE. Reprinted with permission.)

For path $P_3 = (a, G_1, G_3, G_4, G_2, F)$, the Boolean difference chain is $-\nabla(P_3) = (b)(b)(f)(\bar{a} + \bar{b}) = \bar{a}bf$. Since, $(\bar{a}bf) \cdot a = 0$, we do not have a test for an s-a-0 fault.

For path $P_4 = (b, G_1, G_3, G_4, G_2, F)$, the Boolean difference chain is $-\nabla(P_4) = (1)(a)(b)(f)(\bar{a} + \bar{b}) = 0$. This implies that there is no test for the path. If we examine the circuit, we see that the connection from gates G1 to G3 is redundant at the s-a-1 condition.

For path $P_5 = (b, G_3, G_4, G_2, F)$, the Boolean difference chain is $-\nabla(P_5) = \bar{a}f + \bar{b}f$. Thus, the test sequences are $ab = [11, 01]$ and $[11, 10, 00]$.

Test generation using the Boolean difference is a powerful technique. It not only guarantees finding a test sequence if one exists, but it also finds all possible tests. When the Boolean difference is 0, it represents a redundancy in the circuit. One major drawback with this procedure is that it is computationally very expensive. The requirement of very large processing time makes it impractical, even for moderately sized circuits having about 20 to 25 states.

8.4.2 Iterative Logic Array Model

The main difficulty in testing a sequential circuit is due to the presence of feedback lines. If these feedback lines are eliminated, the circuit will become combinational, and the sequential test generation problem reduces to combinational test generation. However, this modified circuit will show the behavior of the original circuit for only a one time step because state values will not be fed back. The behavior of the original circuit can be preserved by using multiple copies of it. The output of the first copy is fed to the second copy, the output of the second copy is fed to the third copy, and so on. Thus, the original sequential circuit is represented by p copies of combinational block (C^r), or an iterative logic circuit in which ith block represents the behavior of the original circuit during the ith time step. This concept is illustrated in Figure 8.6. Figure 8.6(a) shows a generalized sequential circuit, and Figure 8.6(b) shows the iterative combinational model.

The first step in this approach is to select points in the original circuit to cut feedback loops. The cut points should be the places where a delay is expected. In other words, if the delays are inserted between two copies, the behavior of the resulting circuit becomes identical to that of the original circuit. Weights are assigned on each line with the constraint that high fan-out lines should have a high weight. The procedure for deciding the cut points is given as follows [6]:

1. Assign a weight on each primary input equal to its intrinsic weight. The intrinsic weight of a line is given by (number of fan-outs − 1). This intrinsic weight helps to calculate the final weight of a line.

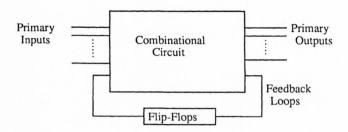

Figure 8.6(a) A generalized sequential circuit.

2. If the weights on all predecessor lines are known, assign a weight on this line as (sum of weights on all predecessor lines + its intrinsic weight).
3. Continue step 2. If all weights are assigned, stop. Otherwise, go to step 4.
4. Calculate a strongly connected component (SCC) for each unweighted line. SCC is a maximal set of lines in which, for each pair (i, j), there is a directed graph connecting i to j. Locate an SCC (A), for which all predecessors are weighted.
5. We want to eliminate minimum-period loops within A. For each block k, its period is the length of the shortest loop containing k. Let B be a subset of all blocks of minimum-period within A. Let $A - B = U$, and U_1 is the subset of nearest successors of B in U. The cut point is the input of an element in U_1, which is fed by a minimal weight line.
6. If the conditions in step 5 are not satisfied, select a block (b) in A which feeds some blocks outside A. The cut point is made on a line connecting b to the remainder of A.

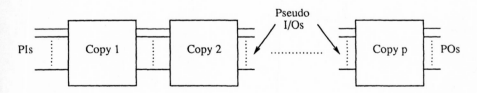

Figure 8.6(b) Iterative logic array model of a sequential circuit.

7. The block following the cut point is assigned a weight (1 + maximum weight assigned so far). Assign weights on all lines that can be assigned after this cut.
8. Repeat steps 4 to 7 until all lines are weighted.

This procedure is best illustrated with an example.

Example 5. Consider the circuit shown in Figure 8.7. First, the intrinsic weights are assigned as:

1	2	3	4	5	6	7	8	9	10	11	12	13	14
0	0	0	0	1	0	1	1	0	0	0	0	2	−1

Based upon these weights, primary inputs 1, 2, 3, and 4 are assigned as weight 0. Line 7 is assigned weight 1 in step 3. At this point, unweighted successors of weighted lines are 5, 6, 8, 9, and 11. In this set, only line 5 has an SCC (5, 6), which has all predecessors weighted. Also, an SCC of 5 is identical to its subset B in step 5; hence, step 6 is used. In step 6, a cut point is made in the loop from block 5 to block 6, marked as X1. Based upon cut X1, lines 6 and 5 are assigned weights of 2 and 3, respectively. At this point, we have an SCC (8, 9, 10, 11, 12, 13). The minimal period lines in this SCC are 11 and 13. The set U is (8, 9, 10) and U_1 is (8). Therefore, we make a cut point X2, where block 13 feeds block 8, and assign weights on lines 8, 9, 10, and 11. Finally, we have an SCC (11, 13). We make a cut point X3, where block 13 feeds block 11, and assign weights to lines 11, 13, and 14.

After cutting the feedback lines, test vectors can be generated by a standard procedure such as the D-algorithm.

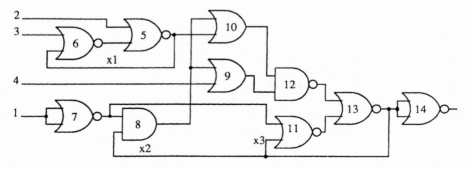

Figure 8.7 An example sequential circuit to illustrate the procedure to eliminate feedback lines. (From [6], © 1971 IEEE. Reprinted with permission.)

8.4.2.1 Application of the D-Algorithm

A fault in the original sequential circuit appears as a multiple fault in the iterative logic array, i.e., the same line appears as a stuck-at fault in each copy of the combinational circuit. The test vectors can be generated by considering the fault multiplicity and also by noticing that pseudoinputs are not controllable and pseudo-outputs are not directly observable.

To obtain a test in P time frames, an array of P-blocks is used. A time frame j is selected during which the fault is sensitized in the jth block. In other words, a D-drive originates from the jth block. Thus, in the jth block, the line under consideration (l_j) is assigned a value D (or $\overline{D}$). A shortest path is chosen from line l_j to the primary outputs and the fault effect is propagated through it. After a successful propagation to the next block, a consistency check is made in the last block. If an inconsistency is discovered, another path is selected for the D-drive. If there is no inconsistency, the D-drive continues until D or $\overline{D}$ reaches a primary output. Once forward propagation is completed, the signal of the blocks whose output are not x (don't-care) are propagated toward the primary inputs. Again, one path is selected for this backward propagation and, if an inconsistency is observed, an alternative selection is made. When this process is completed, a test sequence is obtained.

Throughout the procedure, the faulty line in each block should take a value consistent with its stuck-at value. For example, if a line is s-a-0, in all time frames it should be either 0 or D. Also, special attention should be paid to pseudoinputs and they should be initialized to x. Any other value of a pseudoinput should be considered as an inconsistency.

Note that if the original circuit is asynchronous, we change its characteristics by introducing delays while modeling it as an iterative logic array. Hence, a test sequence obtained by the above procedure should be considered only as a potential test for asynchronous circuits. The test sequence should be simulated to confirm its validity.

Example 6. To illustrate this procedure, let us calculate a test for a line 7 s-a-1 fault in the circuit given in Figure 8.7. Let us choose two time frames and start D-propagation in the second time frame. We will use subscripts 1 and 2 on the line values to show the time frame. To start D-propagation, we assign $\overline{D}$ on line 7_2, which also means 1 on line 1_2. To propagate $\overline{D}$ through block 11_2 (11_2 being D), we assign 0 on 13_1 and 1 on 14_1. It further propagates through 13_2, causing $\overline{D}$ on 13_2, D on 14_2, 0 on 12_2, 1 on 9_2, and 0 on 2_2 and 6_2. This completes D-drive operation.

We need to justify 0 on 13_1 and 6_2. A 0 on 13_1 can be obtained by a 1 on 11_1 or 12_1. A 1 on 11_1 implies a 0 on pseudoinput SI_2, and hence an inconsistency. A 1 on 12_1 can be obtained by a 0 on 9_1 or on 10_1. However, both causes a 0 on 8_1, which implies a 0 on 7_1. Since line 7 is s-a-1, this is an inconsistency.

Backtracking the process, we find 8_2 to propagate the fault effect. To obtain a $\overline{D}$ on 8_2, we assign 1 on 13_1, 0 on 14_1, 11_1 and 12_1, 1 on 7_1, 9_1 and 10_1, 0 on 1_1. This further propagates by a $\overline{D}$ on 9_2 and 0 on 4_2. Finally, we obtain D on 12_2, $\overline{D}$ on 13_2, D on 14_2, 1 on 10_2 and 5_2, 0 on 2_2 and 6_2. This propagation is shown in Figure 8.8. We need to justify 1 on 9_1, 1 on 10_1, and 0 on 6_2. We obtained this by setting 1 on 4_1, 1 on 5_1, 0 on 2_1, 6_1 and 1 on 3_1. Thus, the test sequence with respect to inputs is given as [0011, 10 x 0].

The main advantage of this method is that the test sequences for sequential circuits can be obtained by the direct application of a combinational test method. By adding a few subroutines, an ATPG can be used for sequential test generation. Because, this method is based upon stuck-at faults, the testing is more rigorous than the checking experiment, as discussed in Section 8.3.2. The application of the D-algorithm, however, is costly in terms of processing time and not practical, even for a moderately sized circuit. For small circuits having less than 100 gates, it can be effectively used.

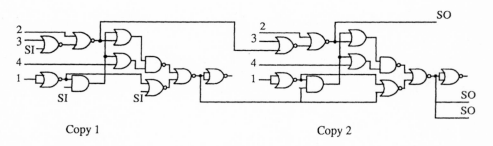

Figure 8.8 Test generation using the D-algorithm. (From [6], © 1971 IEEE. Reprinted with permission.)

8.4.2.2 Application of the Critical Path Tracing

Instead of the D-algorithm, test sequences for sequential circuits can be obtained by the *critical path tracing* method, using an iterative array model. However, because the effect of fault propagation to more than one state variable in one time frame may interact with the previous or subsequent time frame, special attention must be paid during backtracking. The method for combinational circuitry needs to be modified to consider fan-in and fan-out points. Also, this method assumes that the circuit can be reset into a known state that is not affected by faults in the circuit. The modification to combinational critical path tracing is given in Figure 8.9, where CRIPT(x) denotes backtracking from line x using a procedure for combinational circuits, as given in Section 5.3.1. Using this modified procedure

/* Procedure to identify reconvergent gates and fan-out stems */
For each reconvergent gate G
    For each stem s of G
        For $p = 0, 1$
        Determine $I_p(G, s)$
        If $|I_p(G, s)| > 1$ and s is not covered by
            any other stem with respect to G and $I_p(G, s)$
        Flag gate G, associate stem s and $I_p(G, s)$ with G
    If $I_0(G, s) \cap I_1(G, s) \neq \Phi$
    For each stem a_i of G which is reachable from s
    If $I_0(G, a_i) \neq \Phi$ and $I_1(G, a_i) \neq \Phi$
    Include s in the set of p-stems of a_i

/* Procedure EXCRIPT(x) */
1. CRIPT(x)
2. If CRIPT stopped at the output of a flagged gate G
    For every stem s of G
    Perform stem analysis
    If stem s is critical
    EXCRIPT(s)
3. If CRIPT stopped due to self-masking at a stem s which is noncritical
    For every p-stem k associated with the noncritical stem s
    Perform stem analysis
    If k is critical
    EXCRIPT(k)

Figure 8.9 Procedure EXCRIPT to generate test vectors for sequential circuits. (From [7], © 1991 IEEE. Reprinted with permission.)

(called EXCRIPT [7]), test sequences for sequential circuits can be obtained as follows:

1. Simulate the current time frame (jth block) using true value simulation.
2. Insert fault effect at the pseudoinput of the block if the fault effect reached state variables in the preceding time frame.
3. Insert this fault at the fault site and propagate its effect through the current time frame. If it propagates to the primary output, mark it as detected. Update the set of target faults.
4. Backtrack each primary output for every critical line using EXCRIPT. If a fault was not included in steps 2 and 3, mark it as detected.
5. Backtrack each pseudooutput using EXCRIPT. For every critical line with a corresponding fault, include the pseudoinput in the set of target faults if this fault was not processed in steps 2 and 3.

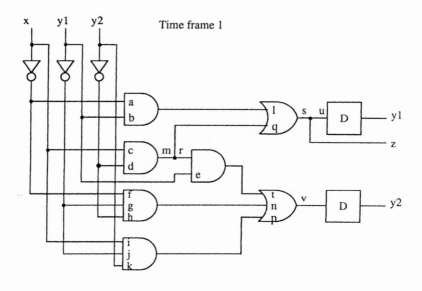

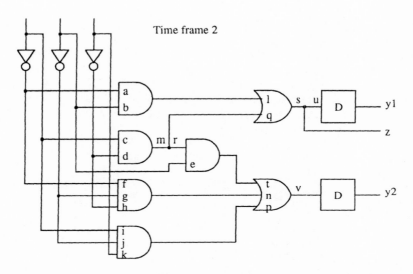

Figure 8.10 Sequential circuit in two time frames for Example 7. (From [7], © 1991 IEEE. Reprinted with permission.)

Example 7. Consider the circuit shown in Figure 8.10. This circuit has one primary input (x), one primary output (z), and two state variables y_1 and y_2. Let the reset state of the circuit be $y_1, y_2 = 10$. Consider the first time frame with input $x = 1$. For this, we obtain $y_1 = 1$, $Y_1 = 1$, $y_2 = 0$, $Y_2 = 1$ and $z = 1$. This detects s-a-0 faults at lines c, d, m, q, s, z. At this time the sets of target faults are {s-a-0 at e,r,t,v through y_2} and {s-a-0 at u through y_1}. Consider another input $x = 1$, for the second time frame, for which $y_1 = y_2 = 1$. Under this condition, faults s-a-0 at e, r, t, v are propagated to Y_1. The faults s-a-1 at a, d, l, m, q, s, z are also detected during the second time frame. After the second time frame, the sets of target faults are {s-a-1 at j, n, p, r, t, v through y_2} and {s-a-1 at u through y_1}.

This method requires less processing time than the D-algorithm. However, it is not an exact method and may not be able to find a test for some detectable faults. Although this method has been useful for synchronous circuits, its extension to asynchronous circuits is not clear. Another limitation of this method is that it assumes a fault-free reset state. This assumption may not be valid in many cases.

8.4.2.3 Cascaded and Interacting Sequential Circuits

In Figure 8.6(a), a sequential circuit is represented by a simple model. In many cases, a more complex model is required. One possibility is that a circuit appears as a cascade of two finite state machines, FSM1 and FSM2 (Figure 8.11). Another possibility is that circuit appears as two mutually interacting finite state machines (Figure 8.12(a)). Under the assumption that both FSMs have a reset state, an iterative logic array model can be used to generate test vectors. In this procedure, the interacting lines are first cut to obtain two individual FSMs. Then, each FSM is represented as an iterative logic array, and test vectors are independently generated for each machine.

For the case of Figure 8.11, to generate a test for FSM1, a test sequence (T_1) is calculated, which generates a value D or $\overline{D}$ at its output. This value propagates through FSM2 by a sequence ($T_{1.2}$) until it reaches the primary outputs. By com-

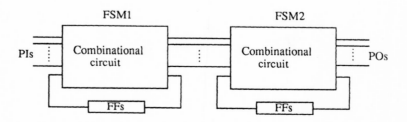

Figure 8.11 Cascaded sequential circuit.

bining $(T_1, T_{1,2})$, a test sequence is obtained that detects a fault in FSM1. To detect a fault in FSM2, a test sequence (T_2) is calculated, which generates a value D or $\overline{D}$ at the primary outputs. The values at the inputs of FSM2, which sensitize the fault, are propagated backward through FSM1 until primary inputs are assigned. Let us say that a sequence $(T_{2,1})$ at the primary input produces a sensitizing vector at the input of FSM2. The sequence (T_2) is concatenated with $(T_{2,1})$ to obtain a sequence that detects a fault in FSM2.

Mutually interacting machines have an additional feedback loop. By cutting this loop, the circuit is converted into that of Figure 8.11. In other words, we can represent the circuit of Figure 8.12(a) by an iterative array in which each block corresponds to Figure 8.11. This is shown in Figure 8.12(b). As discussed above, a test sequence can be obtained for cascaded machines. This sequence is evaluated in different time frames to sensitize a fault in the jth block of Figure 8.12(b) and propagates the fault effect to the output of the pth block.

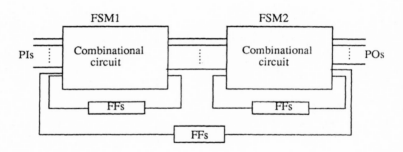

Figure 8.12(a) Interacting sequential circuit.

Few attempts have been made to calculate test sequences for cascaded and interacting sequential circuits. However, it is still an open problem. A practical test generation method for even small cascaded and interacting circuits has not been developed. The observability and controllability is extremely poor in these

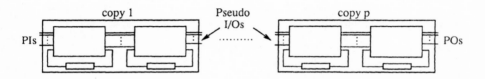

Figure 8.12(b) Iterative array model for interacting sequential circuit.

circuits. Instead of an attempt to calculate a test sequence, a better approach would be to redesign these circuits to incorporate testability features, as discussed in Chapter 10.

8.4.3 Simulation-Based Test Generation

Recent developments and the availability of circuit simulation tools has motivated the development of a simulation-based procedure to identify which input can be used as a test vector. The concept is a directed search of the whole set of input vectors, starting with any one vector. The circuit is simulated and, based upon this simulation, a cost function is computed. This cost function measures *how far a vector is from detecting a fault*. A vector is identified as a test vector for some target fault if the cost function is less than a predefined value [8]. If the computed cost is higher than the predefined value, a change is made in the vector so that the cost is reduced. This procedure is repeated until a vector is identified as a test vector.

To calculate the cost, a weight is computed for each line. This weight represents the effort needed to change the logic value at that line. If, for line i, we denote it as V_i, the average weight for an n-input gate is given by

$$V = \frac{1}{n}\left[\sum_{i=1}^{n} V_i\right] \tag{8.3}$$

For example, for a three-input AND gate, if the inputs are 110, $V_{out} = 0.667$. Based upon these weights, an average input value can be calculated at which the gate output switches from logic 0 to 1 or *vice versa*. This average input value is referred to as a *gate threshold function*. To identify the noise margins, the range for logic 0 is given by $[0, S_0]$, and the range for logic 1 is given by $[S_1, 1]$. Thus, the maximum and minimum threshold of a gate can be calculated if S_0 and S_1 are known. For example, for an AND and OR gate, these bounds are

$$1 - \frac{1 - S_0}{n_{max}} < t_{AND} \leq S_1 \; ; \; S_0 \leq t_{OR} < \frac{S_1}{n_{max}}$$

The threshold function of any Boolean function can be obtained by using primitive gates, i.e., OR, AND, NOR, NAND, and NOT. For example, the threshold function of an XOR gate is

$$T_{XOR}(V_A, V_B) = T_{OR}\left[\frac{T_{AND}\left(\frac{1 - V_B + V_A}{2}\right) + T_{AND}\left(\frac{1 - V_A + V_B}{2}\right)}{2}\right]$$

Where $T_{inv} = [1 - V]$ is the threshold function for an inverter, and T_{AND} and T_{OR} are shown in Figure 8.13(a) and (b), respectively.

To calculate the cost, all circuit elements are leveled, primary outputs being at level 0. A flip-flop output is considered as a pseudo-observation point, and its level is determined by the smallest number of flip-flops on a path from this to a primary output. The cost at level n is given by:

$$(\text{Cost})_n = \min (i \in I_n) \left[\frac{1}{TV_g (i) - TV_f (i)} \right] \tag{8.4}$$

Where I_n is the input to a flip-flop at level n, and $TV_g(i)$ and $TV_f(i)$ are the threshold values at line i in the good and faulty circuits. The cost for an input vector is

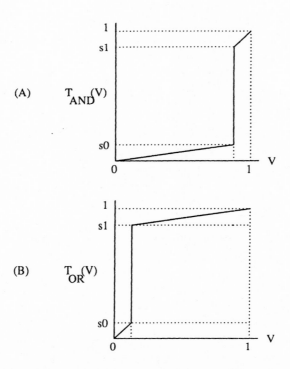

Figure 8.13 Threshold functions (A) for AND gate and (B) for OR gate. (From [8], © 1990 IEEE. Reprinted with permission.)

calculated for each level. The total cost is a weighted sum of all costs. If a fault is not detected by a vector, weighted cost helps to direct the search. The overall test generation procedure is given in Figure 8.14. Note that this is an iterative process and does not guarantee finding a test vector for a given fault. During the minimization of cost function, cost above the threshold function is sometimes obtained, which is at the local minimum. This terminates the search, identifying that the vector is not a test. Thus, a fault is considered undetected.

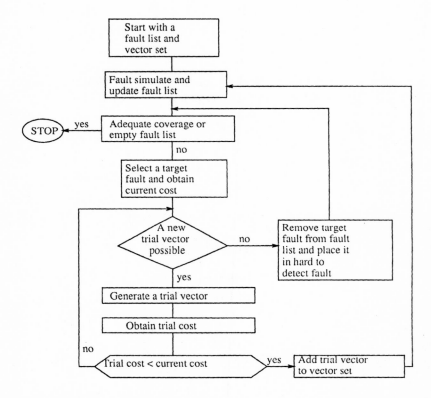

Figure 8.14 Simulation-based test generation procedure. (From [8], © 1990 IEEE. Reprinted with permission.)

The major advantages of this method are that it does not require a combinational circuit model and multiple backtracking. This method can be implemented by an artificial neural network, which can be trained by simulation. This method is highly processor-intensive and, in general, very costly.

8.4.4 Divide and Conquer

The recent efforts in sequential test generation methods are toward the verification of state transition graph methods (discussed in Section 8.3), while using gate-level test generation methods. The sequential test generation problems can be divided into three subproblems [9]:

1. Combinational test generation, assuming present state lines as the primary inputs and the next state lines as the primary outputs;
2. Calculation of *state justification sequence* (J_S), for a fault-free circuit (a sequence that brings the circuit from the reset state to the excitation state through which a given fault can be sensitized);
3. Calculation of *state differentiation sequence* (D_S), for a fault-free circuit (a sequence that can uniquely differentiate between any two states S_i and S_j by producing different primary outputs).

Application of J_S and D_S verifies all state transitions. The J_S and D_S can be calculated by using a binary decision diagram (as discussed in Section 8.3). A fault can be detected if the excitation state in the faulty circuit can be justified, and the fault effect can propagate to the primary outputs by a combinational test vector. If the fault effect reaches the next state lines instead of the primary outputs, an additional state differentiation sequence is required. Note that J_S is a transfer sequence and D_S is similar to a distinguishing sequence.

For step 1, combinational test vectors can be obtained by any standard method such as the D-algorithm or PODEM. For steps 2 and 3, either STG or state *covers* should be extracted from the structural or gate level description of the sequential circuit. Cover extraction has been observed to be significantly faster [9]. In this process, each flip-flop input and the primary outputs are represented as separate cone circuits. A complete or partial sum-of-product representation of the on-off sets (an on-off set comprises the input values which cause a 1/0 at the primary output or the next state lines) for these cones is extracted by the D-algorithm or PODEM. The overall test generation method is given as follows [9]:

1. Extract the on-off sets of the primary output and the next state lines.
2. For the given fault, calculate a combinational test vector.
3. If test vector has an unjustifiable excitation state, calculate a new vector which has different present state than the previous vector. If there is no vector, exit without a test.
4. Calculate fault free J_S for the excitation state. If J_S cannot be found, go to step 3.
5. Simulate J_S for the fault. If fault is detected, generate test sequence directly from J_S. If J_S is valid, go to step 6; otherwise, identify the state for which the edge is faulty and differentiate between the faulty and fault-free state pair.

6. Calculate fault-free D_S for the faulty and fault-free state pair. If D_S is not found, go to step 7; otherwise, combine J_S and D_S to obtain a test.

7. Extract the faulty cover of the primary outputs and next state lines. Simulate the circuit by specifying don't-care values as 1 or 0 in J_S to derive the faulty and fault-free state pairs. Calculate D_S under fault. If no D_S is found and all assignments to don't-care values are not exhausted, go to step 4; otherwise, use a different assignment on don't-care values. Combine J_S and D_S to obtain a test.

This procedure (called STEED) has been implemented and tested on various circuits having a few to more than 200 flip-flops. For a set of sequential benchmark circuits, this has been found to be quite fast, providing close to 100% fault coverage for every circuit.

The sequential test method given above is quite efficient. In [10], this method is further modified to generate J_S and D_S using a behavioral description at the *register transfer level* (RTL), instead of extracting covers from the gate-level description. The algorithms to generate J_S and D_S are given in Figures 8.15(a) and (b). The RTL description of circuits is generally available. The use of this information along with the gate-level description improves the efficiency of the test generation

```
Justify_state (state)
{
    Assign all unknown logic level present state wires a value;
    Store all assignments in binary decision tree;
    While (binary decision tree ≠ Φ) {
    value_set = values on each RTL next state wire;
    flag = justify_value (value_set, 0);
    if (flag) {
        return (justification sequence);
    }
    else {
        /* Justification sequence not found */
        assign last variable on binary decision tree a different value;
    }
    }
    /* No justification sequence obtainable */
    return (justification not possible);
}
```

Figure 8.15(a) Procedure to obtain justification sequence at RTL level. (From [10], © 1990 IEEE. Reprinted with permission.)

```
differentiate()
{
   /*  Given the NS wires that has true/faulty value  */
   if (differentiating input combination exists for NS wire) {
      store differentiation sequence;
      return (TRUE);
   }
   for (each PS wire with true/faulty value) {
      for (each output) {
         for (each path from the PS wire to output) {
            sensitize path;
            simulate assignments;
            justify all values on intermediate wires;
            if (all values justified) {
               store differentiation sequence;
               return (TRUE);
            }
         }
      }
   }
   /*  True/faulty value cannot be propagated to output  */
   for (each PS wire with true/faulty value) {
      for (each NS wire) {
         for (each possible path form the PS to NS) {
            sensitize path;
            simulate assignments;
            justify all values on intermediate wires;
            if (all values justified) {
               flag  =  differentiate (new_values);
               if (flag) {
                  make up differentiation sequence;
                  return (TRUE);
               }
            }
         }
      }
   }
   /*  Cannot differentiate this state pair  */
   return (FALSE);
}
```

Figure 8.15(b) Procedure to obtain differentiation sequence at RTL level. (From [10], © 1990 IEEE. Reprinted with permission.)

program (the program STEED has been modified; new version is called ELEK-TRA). The test vectors providing almost 100% fault coverage reportedly can be obtained for even large, complex circuits such as digital signal processing chips.

8.5 SYNTHESIS FOR TESTABILITY

The above methods calculate test sequences for sequential circuits that have no testability features. In most of the cases, these methods are successful, but there are instances when a test is not found. Circuit redundancies may cause such situations. Assuming that the combinational part of the circuit is nonredundant (it can be achieved by using simple logic optimization methods), sequential redundancies should be identified and eliminated to achieve 100% testability. Three possible *sequentially redundant faults* (SRFs) have been identified:

1. *Invalid state fault*: A fault which corrupts the edge of an invalid state;
2. *Isomorphic fault*: A fault which changes the encoding of states;
3. *Equivalent state fault*: A fault which creates equivalent states in the STG.

There are two possibilities to avoid these redundancies to obtain complete testability. One possibility is to use appropriate don't-care sets during synthesis and another possibility is to restrict the implementation by specific constraints during synthesis. In [11] and [12], synthesis procedures have been given to obtain complete testability. Some observations regarding the testability of a sequential circuit are summarized as follows:

1. Given a reduced sequential circuit with 2^n states, where n is the number of flip-flops, if the combinational logic is prime and nonredundant and is implemented in two-level form or algebraically factored multilevel form, the circuit is fully testable for all single stuck-at faults in the combinational logic.
2. Given a reduced sequential circuit of either Moore or Mealy form with $N_s \leq 2^n$, where n is the number of flip-flops, if the combinational logic is prime and nonredundant with respect to the invalid state don't-care set, all single stuck-at faults on the primary inputs or the present state lines and all single and multiple stuck-at faults on the primary outputs or the next state lines are testable.
3. Given a Moore or Mealy machine's logic-level implementation with n flip-flops, if (1) combinational logic blocks and next state lines are nonredundant (2) the machine has a reset state and it is R-reachable, i.e., all 2^n states are reachable from the reset state, (3) the state codes of the states with the same output has a distance of at least two from each other, the machine is fully testable.
4. If each of the input labels in at least one differentiating sequence of at least one faulty and fault-free state pair produced by a fault are made fault-effec

disjoint from the input label, the perturbation of which caused the faulty and fault-free state pair, the resulting circuit is fully testable.

Points 1 and 2 indicate detection of some faults in any sequential circuit. Observation 1 dictates 100% stuck-at fault detection in the combinational part if a fault cannot create a new state.

Observations 3 and 4 indicate that with careful design and synthesis a completely testable sequential circuit can be obtained. Point 3 gives the constraints on state assignment that any pair of states which cannot be distinguished by a single vector sequence should be given codes at least a distance of two apart. Point 4 suggests implementing a circuit for which a particular type of differentiating sequence can be obtained. The circuit should have the following properties:

1. An input test vector should exist for every single stuck-at fault in the network that lies in the on-off sets and outside the don't-care set.
2. On application of this test vector, at least one of the output values that differ between the true and faulty circuits will not correspond to a don't-care output condition.

A synthesis procedure that guarantees a differentiating sequence for valid-invalid state pair is given in Figure 8.16. This procedure provides a circuit-satisfying observation 4, and hence is fully testable. A specific use of don't-care sets provides optimized logic, but costs in terms of processor time; constrained synthesis requires small processor time, but does not provide optimized implementation in terms of hardware. However, a fully testable sequential circuit can be obtained by a little care in design and synthesis.

8.6 SUMMARY

Sequential circuit testing is significantly more complex than combinational circuit testing. The major difficulty is due to the memory elements and feedback loops. We discussed a method based upon a binary decision diagram to find initializing sequences for any sequential circuit. We gave a functional testing scheme based upon verification of state table or state transition graph. While such testing reduces the effort of a test engineer, it does not provide high fault coverage.

Gate-level test methods are discussed to detect stuck-at faults in the sequential circuits. Extensions of the combinational Boolean difference method, D-algorithm, and critical path tracing methods are given. However, these methods are highly processor-intensive and can only be used for small circuits. A simulation-based test generation method is given, which is also processor-intensive. A test method based upon STG verification using gate-level test vectors is given. This method divides the sequential test problem into multiple subproblems, and has been found to be very efficient.

```
eliminate sequential redundant faults (S);
{
    iter = 1;
    do {
        if (iter = 1) G = extract-STG(S);
        else G = extract-STG(S");
        foreach (valid state q ∈ G) {
            find all valid states (v₁, . . . , vₘ) ≡ q;
            find all invalid states (iv₁, . . . , ivₙ) ≡ q;
            DC₁: fan-in (q) = (q, v₁, . . . , vₘ, iv₁, . . . , ivₙ);
            find all input-labels i_{q·s} differentiating q and s ≠ q;
            DC₂; fan-in (q) = (q, s)&& n(i_{q·s}, q) = n(i_{q·s}, s)&&
            o(i_{q·s}, q) = o(i_{q·s}, s);
        }
        S' = optimize (S, DC₁, DC₂);
        IV = extract-invalid-states (S');
        S'' = optimize (S', DC_{IV});
        iter = iter + 1;
    } while (S ≠ S");
}
```

Figure 8.16 Synthesis procedure for which a specific differentiating sequence always exists. (From [12], © 1991 IEEE. Reprinted with permission.)

Synthesis of sequential circuits is discussed in consideration of testability. Some conditions have been identified to assign states during synthesis such that 100% testability can be obtained in the final circuit. Finally, a synthesis procedure is given, which guarantees the presence of a specific differentiating sequence, and hence a fully testable sequential circuit.

PROBLEMS

1.   Consider a FSM, the state table of which is as follows:

Present State	Next State and Output	
	Input 0	Input 1
A	C,1	B,0
B	C,0	B,1
C	D,1	C,1
D	A,1	C,0

Calculate (a) a homing sequence, (b) a distinguishing sequence, and (c) a synchronizing sequence for this machine.

2. Suppose that the FSM given in Problem 1 can be reset to state A. Obtain a minimal length test sequence to verify all transitions or edges in its STG.

3. Use the D-algorithm to obtain a test sequence for the output of a gate G1 s-a-1 fault in the circuit of Figure 8.5.

4. Use an iterative logic array model and EXCRIPT to obtain tests for various lines in the circuit of Figure 8.5.

5. The FSM given in Problem 1 has been implemented by (a) four flip-flops, (b) two flip-flops. Give a gate-level diagram for both circuits. Which circuit is more testable? Assign a code to all states in the four flip-flops implementation to make it 100% testable.

REFERENCES

1. Z. Kohavi and P. Lavalee, "Design of Sequential Machines with Fault Detection Capabilities," *IEEE Trans. Elect. Comp.*, **16**, Aug. 1967, pp. 473–484.

2. T.N. Hibbard, "Least Upper Bounds on Minimal Terminal State Experiments for Two Classes of Sequential Machines," *J. ACM*, **8**, Oct. 1961, pp. 601–612.

3. C.R. Kime, "An Organization for Checking Experiments on Sequential Circuits," *IEEE Trans. Elect. Comp.*, **15**, Feb. 1966, pp. 113–115.

4. D.E. Farmer, "Algorithms for Designing Fault Detection Experiments for Sequential Machines," *IEEE Trans. Comp.*, **22**, Feb. 1973, pp. 159–167.

5. M.Y. Hsiao and D.K. Chia, "Boolean Difference for Fault Detection in Asynchronous Sequential Machines," *IEEE Trans. Comp.*, **20**, Nov. 1971, pp. 1356–1361.

6. G.R. Putzolu and J.P. Roth, "A Heuristic Algorithm for the Testing of Asynchronous Circuits," *IEEE Trans. Comp.*, **20**, June 1971, pp. 639–647.

7. P. Menon, Y. Levendel, and M. Abramovici, "SCRIPT: A Critical Path Tracing Algorithm for Synchronous Sequential Circuits," *IEEE Trans. CAD*, **10**(6), June 1991, pp. 738–747.

8. K.T. Cheng, V.D. Agrawal, and E.S. Kuh, "A Simulation Based Method for Generating Tests for Sequential Circuits," *IEEE Trans. Comp.*, **39**, Dec. 1990, pp. 1456–1462.

9. A. Ghosh, S. Devadas, and A.R. Newton, "Test Generation and Verification for Highly Sequential Circuits," *IEEE Trans. CAD*, **10**(5), May 1991, pp. 652–667.

10. A. Ghosh, S. Devadas, and A.R. Newton, "Sequential Test Generation at the Register Transfer and Logic Level," *Proc. Design Auto. Conf.*, 1990, pp. 580–586.

11. S. Devadas, H.T. Ma, A.R. Newton, and A.S. Vincentelli, "A Synthesis and Optimization Procedure for Fully and Easily Testable Sequential Machines," *IEEE Trans. CAD*, **8**(10), Oct. 1989, pp. 1100–1107.

12. S. Devadas and K. Keutzer, "A Unified Approach to the Synthesis of Fully Testable Sequential Machines," *IEEE Trans. CAD*, **10**(1), Jan. 1991, pp. 39–50.

Chapter 9
Microprocessor Testing

9.1 INTRODUCTION

A microprocessor is the brain of a computer system. In general, it is a multiple, interacting, cascaded sequential circuit. With the increasing word size of a microprocessor, the number of states becomes so large that only a small fraction of all faults can be tested. Secondly, almost all internal nodes in a control unit are not directly controllable and observable. To find a checking sequence to verify the state transition graph of the control unit is extremely difficult. Detection of all stuck-at faults is almost impossible.

To overcome this complexity, some simplifying assumptions are used to develop a test method for the control unit. As discussed in Chapter 2, functional fault models are used for microprocessor testing instead of the conventional stuck-at fault model. Using functional fault models, some techniques have been developed to test the control unit. Another possibility is to use *scan design* or *built-in self-test* (BIST) methods to make the control unit testable. The detailed discussions on scan design and BIST are given in Chapter 10. In this chapter, we will describe the microprocessor testing methods, based upon functional fault models, which do not require extra hardware for testing.

9.2 MICROPROCESSOR DESCRIPTION AND TESTING

For testing any circuit, some form of the circuit description is essential. In the case of a microprocessor, the functionality and operation is described in several different ways. At a higher level, normal representation of a microprocessor includes two sections, i.e., the *data path* and the *control unit*. The data path can be considered as a simple Moore or Mealy synchronous sequential circuit, and can be efficiently tested by the methods discussed in the previous chapter. The control unit is an

---

interacting, cascaded sequential circuit. These circuits are the most complex and present a real challenge in testing. Although some attempts have been made to develop a test method for the control unit, as we will discuss in this chapter, it is still an open problem.

At the lowest level of representation, like any other circuit, a microprocessor can be described by its actual layout or transistor-level schematics. Theoretically, if transistor level schematics are available, each individual block (i.e., on-chip memory, decoding logic, on-chip PLA, etc.) can be identified and tested independently, using the methods discussed in Chapters 5 to 8. However, due to the size of present-day microprocessors, such testing is so cumbersome that it is practically infeasible. More importantly, transistor-level schematics generally considered proprietary information and are not available to test engineers or microprocessor users.

Another popular description of a microprocessor is its functionality through the instruction set. Each instruction specifies a data manipulation, control, or data transfer operation among the microprocessor registers. This register-transfer level information is available to the user. Hence, from the user's perspective, verification of the complete instruction set is sufficient to determine whether a microprocessor has a fault. The details of this method are given in the next section.

9.3 INSTRUCTION SET VERIFICATION

A very systematic study of verifying the instruction set was reported by Robach and Saucier [1], Thatte and Abraham [2], and Brahme and Abraham [3]. The approach in [1] is close to state table verification. More precisely, a state transition graph has been extracted and enumerated for each instruction. In [2], a generalized graph is developed to represent the overall functionality of the microprocessor. Then, the microprocessor is divided into several blocks. The functionality of individual logic blocks has been considered, while the instructions are considered at the machine level. The functionality of a logic block is again considered in [3], and microinstruction-level test programs are developed.

9.3.1 Machine-Level Verification

The overall functionality of a microprocessor can be given by a system graph, called the S-graph [2]. This graph has *one sink node* (OUT) and *one source node* (IN), representing external connections to the microprocessor. Each register (R_i) is represented as a node, while each instruction is represented as a labeled directed edge between two nodes if the instruction causes a data transfer between those two registers. An example S-graph for a general-purpose microprocessor is shown in Figure 9.1, the description of registers and instructions is given in Table 9.1. All instructions are categorized in three classes as shown in Table 9.1: (a) *branch,*

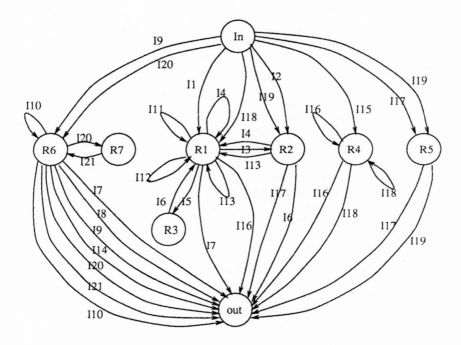

Figure 9.1 S-graph of a general-purpose microprocessor. (From [2], © 1980 IEEE. Reprinted with permission.)

denoted as B; (2) *data manipulation*, denoted as M; and (3) *data transfer*, denoted as T.

To facilitate the test, the whole microprocessor is divided into several blocks with known functionality and tests are developed to verify correct instruction execution for each block. Actual implementation or gate-level schematics of these blocks are not necessarily known. The major functional blocks may be the instruction decoding and sequencing unit, register decoding module, data storage unit, data transfer unit, and data manipulation unit. Only functional faults are considered in each block and some simplified assumptions are used during test generation.

Instruction Decoding and Sequencing Unit

The instruction decoding and sequencing unit is considered faulty if:

1. An instruction does not execute. The special case is that no instruction executes, denoted as $f(I_j/\phi)$.

Table 9.1
Description of registers and instructions in Figure 9.1.
(From [2], © 1980 IEEE. Reprinted with permission.)

Description of registers in Figure 9.1:
R1: Accumulator
R2: General-purpose register
R3: Scratch-pad register
R4: Stack pointer to the top of a LIFO stack in main memory
R5: Address buffer register to store the address of operand
R6: Program counter
R7: Subroutine register to save return address

Description of instruction in Figure 9.1:
I1: Load register R1 from the main memory using immediate addressing (T)
I2: Load register R2 from the main memory using immediate addressing (T)
I3: Transfer the contents of R1 to R2 (T)
I4: Add the contents of R1 and R2 and store the result in R1 (M)
I5: Transfer the contents of R1 to R3 (T)
I6: Transfer the contents of R3 to R1 (T)
I7: Store R1 into the main memory using implied addressing (T)
I8: Store R2 into the main memory using implied addressing (T)
I9: Jump instruction (B)
I10: Skip if the content of R1 is zero (B)
I11: Left shift R1 by one bit (M)
I12: Complement bitwise the contents of R1 (M)
I13: Logical AND the contents of R1 and R2, and store the result in R1 (M)
I14: No operation instruction (B)
I15: Load stack pointer R4 from main memory using immediate addressing (T)
I16: Push R1 on LIFO stack maintained in the main memory (T)
I17: Store R2 into the main memory using direct addressing (T)
I18: Pop the top of the LIFO stack and store it in R1 (T)
I19: Load R2 from the main memory using direct addressing (T)
I20: Jump to subroutine and save return address in R7 (B)
I21: Return from subroutine (B)

2. Instead of a specified instruction, another instruction executes. It is denoted as $f(I_j/I_k)$.
3. An extra instruction is executed in addition to the specified instruction. It is denoted as $f(I_j/I_j + I_k)$.

For these faults in the instruction decoding and sequencing unit, the simplifying assumptions are [3] as follows:

1. If either $f(I_j/I_k)$ or $f(I_j/I_j + I_k)$ is present, I_k will execute correctly.
2. If $f(I_j/I_k) \cdot f(I_j/I_j + I_k)$ or $f(I_j/\phi)$ is present, $f(I_q/I_j)$ and/or $f(I_q/I_q + I_j)$ is no present.

An ordered sequence of tests is applied to detect: (1) $f(I_j/\phi)$ faults; (2) $f(I_j/I_k)$ faults, and (3) $f(I_j/I_j + I_k)$ faults. The test generation procedure is given in Figure 9.2.

Register Decoding Unit

The register decoding unit is considered faulty if:

1. A register is not accessed. The special case is that no register is accessed, denoted as $f_D(R_i) = \{\phi\}$, where f_D represents a register mapping function. In this situation, the contents of a register do not change after writing, and when it is retrieved an all 1 or all 0 pattern is obtained. The fault-free condition is denoted as $f_D(R_i) = \{R_i\}$ for every $R_i \in R$, where R is the set of registers.

```
/* Test generation for f(Iⱼ/φ) */
STEP 1: Store OPERAND 1 in D(Iⱼ) and proper operand(s) in S(Iⱼ)
    such that when Iⱼ is executed it produces RESULT 1 in
    D(Iⱼ) and RESULT 1 ≠ OPERAND 1.
STEP 2: READ D(Iⱼ). /* expected output is OPERAND 1 */
STEP 3: Execute Iⱼ.
STEP 4: Read D(Iⱼ). /* expected output is RESULT 1 */

/* Test generation for f(Iⱼ/Iₖ) */
STEP 1: Store OPERAND 1 in S(Iⱼ) and OPERAND 2 in S(Iₖ)
    such that OPERAND 1 ≠ OPERAND 2.
STEP 2: For i = 1 to k do
    Execute Iⱼ
    READ D(Iⱼ) /* expected output is OPERAND 1 */
    continue
STEP 3: Execute Iₖ.
STEP 4: READ D(Iₖ) /* expected output is OPERAND 2 */

/* Test generation for f(Iⱼ/Iⱼ + Iₖ) */
STEP 1: Store OPERAND 1 in S(Iₖ) and OPERAND 2 in D(Iₖ)
    such that OPERAND 1 ≠ OPERAND 2.
STEP 2: READ S(Iₖ) /* expected output is OPERAND 1 */
STEP 3: Execute Iⱼ.
STEP 4: READ D(Iₖ) /* expected output is OPERAND 2 */
```

Figure 9.2 Test generation procedure for instruction decoding and sequencing unit. (From [2], © 1980 IEEE. Reprinted with permission.)

2. In addition to the specified register, an extra register is accessed. This situation assumes $f_D(R_i) \neq \{R_i\}$ for every $R_i \in R$.

The procedure to detect these faults in the register decoding unit is given in Figure 9.3. The procedure uses two data structures, a queue of registers Q, and a set of registers A. The set A is augmented progressively by removing one register from Q and putting it into A. The test ensures that all registers have disjoint image sets under the register mapping function f_D.

Data Storage and Data Transfer Units

Data storage and data transfer units can be tested by standard memory test procedures as discussed in Chapter 7. However, note that, in addition to general-purpose registers (considered as on-chip RAM), the data storage unit also has some special-purpose registers. A simplifying assumption is that any register may have a stuck-at fault, a transfer line may have a stuck-at fault, or two transfer lines may be coupled (either logically or electrically) with each other. Under these simplifying assumptions, the data storage and data transfer functions can be tested by the procedure given in Figure 9.4. This procedure transfers data among various registers

/* Test generation for register decoding function */
STEP 1: Initialize the queue Q with all the registers such that R_i
lies ahead of R_j iff $I(R_i) \leq I(R_j)$.
Initialize the set A to zero.
STEP 2: A ← register at the front of Q; Update Q.
STEP 3: REPEAT UNTIL Q = empty
(a) Generate instructions to write each register R_i of set A
with data ONE, and the register R_j at the front of Q with
data ZERO; /*instructions of the corresponding WRITE (R_i)
and WRITE (R_j) need to be generated */
(b) Generate instructions to read each register R_i of set A,
such that R_a will be read before R_b iff
$I(R_a) \leq I(R_b)$; /*instructions of the corresponding
READ (R_i) need to be generated */
(c) Generate instructions to read register R_j at the front of Q
(d) A ← A ∪ {register at the front of Q}.
(e) Update Q
STEP 4: Repeat steps 1, 2, and 3 with complementary data.

Figure 9.3 Test generation procedure for register decoding function. (From [2] © 1980 IEEE. Reprinted with permission.)

/* Test generation for any register transfer path T_{j1} to T_{jk} */
/* and registers D_{j1} to D_{jk-1} */
/* width of a register is 8-bits */
I_{j1} with data 11111111 ; /*write D_{j1} with data ONE */
I_{j2} ; I_{j3} ; . . . ; I_{jk} ; /*expected output ONE */
REPEAT I_{j1} to I_{jk} with following data
    (a) 11110000
    (b) 11001100
    (c) 10101010
/* Test for transfer path from ALU to register R_i */
STEP 1: Execute I_1; I_2 with data 11111111, 00000000
    Execute I_4; I_7,
    REPEAT above instructions with following data
      (a) 11110000, 00000000
      (b) 11001100, 00000000
      (c) 10101010, 00000000
      (d) 00000000, 00000000
      (e) 00001111, 00000000
      (f) 00110011, 00000000
      (g) 01010101, 00000000
STEP 2: Repeat step 1 for every R_i and R_j for every
    instruction of class M.
/* Test for transfer path connecting R_1 and R_2 to ALU */
STEP 1: Execute I_1 with 00000001; I_2 with 00000000;
    Execute I_4; I_7;
    REPEAT above instructions with data in I_1 being
    00000010, 00000100, 00001000, 00010000, 00100000, 01000000, 10000000
STEP 2: Execute I_1 with 00000000; I_2 with 00000001
    Execute I_4; I_7;
STEP 3: Repeat steps 1 and 2 for every R_i and R_j for every
    instruction of class M.
/* Test for transfer path associated with class B instructions */
STEP 1: Execute I_9 with jump address 00000000 00000000
    REPEAT with jump addresses
      (a) 00000000 11111111
      (b) 00001111 00001111
      (c) 00110011 00110011
      (d) 01010101 01010101
    REPEAT with complementary set of jump addresses
STEP 2: Repeat step 1 for every instruction of class M.

Figure 9.4 Test procedure for data storage and data transfer paths. (From [4] © 1980 IEEE. Reprinted with permission.)

to verify the correctness of the data transfer unit. It also writes and reads different data patterns into different registers to verify the correctness of the storage capability and to detect coupling faults.

One drawback of this method is that it requires test programs at the machine instruction level. Second, the length of the test subroutines may become quite large. For example, the test sequence for register decoding function is on the order of N^3, where N is the number of registers. The length of the test sequence for instruction decoding and sequencing unit is worse. Even if we assume that the instruction labels do not exceed two, the length of the test sequence for instruction decoding is on the order of n^2, where n is the number of instructions. For present-day microprocessors, n easily may be close to 200, implying a large test subroutine. Generation of such large test programs at the machine instruction level may be cumbersome. Despite these drawbacks, this method has been widely used. One major advantage of this method is that the circuit implementation details are not required for testing. It is a very elaborate process and has been proved successful on various commercial microprocessors [4]. To facilitate testing, various test program libraries have been developed for different microprocessors based on this method.

9.3.2 Microinstruction-Level Verification

Instead of testing microprocessors by the programs at machine instruction level, programs at microinstruction level can be used. This approach has been adopted in [3]. Furthermore, the concept of incremental testing can be used to avoid the extremely large test subroutines. In incremental testing, the functionality of a microprocessor is described in terms of layers, where each layer is a set of some instructions. These layers are identified such that any instruction at the ith level layer can be verified by using the instructions at lower level layers. Testing starts with the zeroth level layer and continues until a fault is detected or all layers are tested. For example, assume that by some test method we established that the READ instructions execute correctly. All other instructions can now be tested by loading the codewords into the registers, followed by executing the instruction and reading all the registers.

In [3], all instructions are considered in composed form, made up of a sequence of microinstructions. For example, instruction I_i is equivalent to $[m_1, m_2, \ldots, m_{k,i}]$. Also, each microinstruction m_i is considered as a set of micro-orders $[\mu_{i.1}, \mu_{i.2}, \ldots, \mu_{i.q}]$, which execute in parallel. A complete set of micro-orders can be constructed from the instruction set of microprocessor and divided into three to four categories. For example, the set of micro-orders for Motorola 68000 microprocessor is given in Table 9.2. This set of micro-orders also is divided into three categories: (1) type 0 micro-orders represent one register operation; (2) type 1

Table 9.2
Set of micro-orders for Motorola 68000 microprocessor.
(From [3], © 1984 IEEE. Reprinted with permission.)

Category	Micro-orders
Type 0	SWAP, CLR, NOT, NEG, BTST, BSET, BCLR, BCHG, SHIFTLEFT, SHIFTRIGHT
Type 1	MOV, AND, OR, EOR, EXG
Type 2	ADD, MUL, SUB, DIV

represent a logic operation or operation on two registers; and (3) type 2 micro-orders represent an arithmetic operation.

To generate tests, some simplifying assumptions are used:

1. Two registers are considered equivalent with respect to an instruction set, if and only if any instruction can use one register instead of the other for its operand.
2. Two instructions I_i and I_j are considered equivalent if $I_i = I_j + $ NOP.
3. An instruction is considered the inverse of another instruction if the composition of both instructions is equivalent to NOP with or without memory writes.
4. Two micro-orders μ_i and μ_j are considered independent, if and only if $S(\mu_i) \cap D(\mu_j) = \phi$ and $S(\mu_j) \cap D(\mu_i) = \phi$, where S is the source register and D is the destination register.
5. If a micro-order μ_i is part of an instruction I_j, then μ_i is redundant with respect to I_j, if I_j is equivalent to $I_j - \mu_i$.

For instruction sequencing unit, we can now consider faults at microinstruction level. A fault is considered a simple fault, if at most one extra micro-order is active, while any number of micro-orders may remain inactive. A linked fault is a sequence of simple faults. Note that to detect a linked fault, it is sufficient to detect the presence of any of the component simple faults. A microprocessor instruction can be considered faulty under the following conditions:

1. One or more micro-orders of an instruction are not working, implying that the instruction is not executed completely. It is denoted as $f(I - \delta^-)$.
2. In addition to the specified micro-orders, some extra micro-orders are executed with an instruction. It is denoted as $f(I + \delta^+)$.
3. Some specified micro-orders are not active, while some unspecified micro-orders are active with an instruction. It is denoted as $f(I + \delta^+ - \delta^-)$.

Note that these conditions are very similar to the conditions discussed in Section 9.3.1. The only difference is that here a missing or extra micro-order is considered, while in the previous section a missing or extra instruction is considered.

Now, for incremental testing, a zeroth level layer is identified consisting of a minimum number of instructions. These instructions are tested for $f(I - \delta^-)$ faults, $f(I + \delta^+)$ faults. Once these instructions are tested, these can be used to test the next layer. This procedure continues until the complete instruction set is tested.

Example 1. Let us consider a Motorola 68000 microprocessor. Consider a zeroth level layer consisting of only three instructions: (1) MOV a, R_i; load register R_i with contents of memory location a, (2) CMP R_i, R_j; compare contents of register R_i and R_j, and set the Z bit of the status register if $R_i = R_j$, and (3) BEQ a; if the Z bit is set, branch to location a. The $f(I - \delta^-)$ faults in these three instructions can be detected by a subroutine given in Figure 9.5.

```
/* Procedure to detect δ⁻ faults in zeroth layer */
        MOV #0,R1
        BEQ a           /*branch #1
        BRA error
a:      MOV #1,R1
        BEQ error       /*branch #2
b:      MOV #0,R1
        BEQ c           /*branch #3
        BRA error
c:      MOV #1,R2
        MOV #0,R1
        CMP R1,R2
        BEQ error       /*branch #4
        CMP R1,R2
        BEQ error       /*branch #5
        MOV #0,R2
        CMP R1,R2
        BEQ d           /*branch #6
        BRA error
d:      MOV #1,R1
        MOV #1,R2
        CMP R1,R2
        BEQ success     /*branch #7
        BRA error
error: write error
success:    NOP
```

Figure 9.5 Procedure to detect $f(I - \delta^-)$ faults in zeroth layer for Example 1. (From [3], © 1984 IEEE. Reprinted with permission.)

Note that at this time we cannot detect $f(I + \delta^+)$ faults until we test the READ instruction. A READ instruction is considered fault-free if the data read are correct and there are no $f(I + \delta^+)$ faults. The faults of type 0, 1, and 2 in the READ register instruction can be detected by the subroutine given in Figure 9.6.

After verifying the READ instruction, all other instructions can be tested. A simple procedure to detect faults in instruction execution process is given in Figure 9.7, where codewords for different registers are given in Figure 9.8. Thus, the complete testing can be summarized as follows [3]:

Step 1: Execute the subroutine given in Figure 9.5. It detects $f(I - \delta^-)$ faults.

Step 2: Execute the subroutine given in Figure 9.6. It detects a fault in the READ instruction.

```
/* Test for type 0 faults in READ instruction */
for all i ∈ R and for all j ∈ S₀ do {
   READ(i)
   {READ(j) M*(K + 1) times}
   READ(i)
   READ(j)
}

/* Test for type 1 faults in READ instruction */
for all i ∈ R and for all (j·k) ∈ S₁ do {
   READ(i)
   READ(j)
   {READ(k) M*(K + 1) times}
   READ(i)
   READ(j)
}

/* Test for type 2 faults in READ instruction */
for all i ∈ R and for all (j, k, l) ∈ S₂ do {
   READ(i)
   READ(j)
   {READ(k) M*(K + 1) times}
   {READ(l) M*(K + 1) times}
   {READ(k); READ(l) M*(K + 1) times}
   READ(i)
   READ(j)
```

Figure 9.6. Procedure to detect faults in READ instruction for Example 1. (From [3] © 1984 IEEE. Reprinted with permission.)

```
/* Test for LOAD instruction */
for all i ∈ R do {
  for all j ∈ R do {
    READ(j)
    MOV dᵢ, Rᵢ
    READ(j)
  }
}

/* Tests for simple faults in instruction execution */
for all Iᵢ ∈ I do {
    Load the registers with code words
    Execute Iᵢ
    Read all internal registers
}
```

Figure 9.7 Procedure to detect faults in instruction execution process for Example 1. (From [3] © 1984 IEEE. Reprinted with permission.)

Step 3: Execute the subroutine given in Figure 9.7. The first part of this subroutine detects fault in the LOAD instruction and the second part verifies the rest of the instructions.

This method of testing a microprocessor is very comprehensive and widely used. This method also considers a microprocessor as a black box and works only through its instruction set. A test engineer does not need to know the actual implementation of the microprocessor. Another advantage is that this method can be implemented as a self-test, and hence the requirement of an external tester can be eliminated. Although this method is useful for hard-wired control units, it is equally useful for microprogrammed microprocessors. The length of the test program in this method is on the order of $(n \cdot N + N^4)$, where n is the number of instructions and N is the number of registers.

9.4 BIT-SLICED MICROPROCESSORS

The bit-sliced microprogrammed processors can be tested much more efficiently than the method of instruction set verification [5]. In the bit-sliced system, a base unit is used to perform a set of operations on n-bit operands. By interconnecting N base units, generally in an iterative array form, the same operations are performed on nM-bit operands. This helps in developing a family of computers that have different word size, but use the same software. In [5], a simple one-bit micro-

Register	Code Pattern
D0	1111 1101 1111 1111 1111 1111 1111 1111
D1	1111 1110 1111 1111 1111 1111 1111 1111
D2	1111 1111 0111 1111 1111 1111 1111 1111
D3	1111 1111 1011 1111 1111 1111 1111 1111
D4	1111 1111 1101 1111 1111 1111 1111 1111
D5	1111 1111 1110 1111 1111 1111 1111 1111
D6	1111 1111 1111 0111 1111 1111 1111 1111
D7	1111 1111 1111 1011 1111 1111 1111 1111
A0	1111 1111 1111 1101 1111 1111 1111 1111
A1	1111 1111 1111 1110 1111 1111 1111 1111
A2	1111 1111 1111 1111 0111 1111 1111 1111
A3	1111 1111 1111 1111 1011 1111 1111 1111
A4	1111 1111 1111 1111 1101 1111 1111 1111
A5	1111 1111 1111 1111 1110 1111 1111 1111
A6	1111 1111 1111 1111 1111 0111 1111 1111
USP	1111 1111 1111 1111 1111 1011 1111 1111
SSP	1111 1111 1111 1111 1111 1101 1111 1111

Figure 9.8 Codeword assignment for Motorola 68000 microprocessor for Example 1. (From [3] © 1984 IEEE. Reprinted with permission.)

processor cell, equivalent to AMD 2901 slice, is used to illustrate the testing procedure. However, the method is quite general and can be used for any bit-sliced system.

9.4.1 Testing of One-Bit Slice

The internal structure of a generalized one-bit microprocessor cell is shown in Figure 9.9(a). The microinstruction control fields for this slice are given in Figure 9.9(b). Based upon this description, test vectors for each block in Figure 9.9(a) can be calculated to detect a fault that changes the function of the block. To calculate the test vectors for a combinational block, any method given in Chapter 5 can be used. To calculate the test vectors for sequential blocks, the checking sequence approach can be used, as given in Chapter 8. Once test sets for individual blocks are obtained, these can be concatenated to obtain the complete set for the slice.

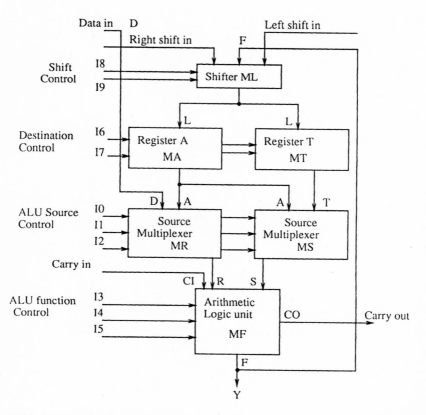

Figure 9.9(a) Architecture of a generalized one-bit slice processor. (From [5] © 1981 IEEE. Reprinted with permission.)

Note that the test output is observable only at the data out line (Y) or at the carry out line (CO).

The first block in Figure 9.9(a), shifter ML has five inputs; hence, an exhaustive test sequence for this block is 32 vectors long. During the application of these 32 vectors, other inputs to the slice are kept at a constant value, chosen such that the output of block ML is observed at Y. For example, in Figure 9.9(a), by setting I_6 and I_7 to 0, and $(I_5, I_4, I_3, I_2, I_1, I_0, CI)$ to $(1, 1, 0, 0, 1, 0, x)$, the output of block ML is observed at Y. Similarly, other combinational blocks MR, MS, and MF can be tested by 66 vectors. The blocks MA and MT are sequential in nature; a sequence of 16 vectors can test these blocks. Thus, a test sequence of length 32 + 66 + 16 = 114 vectors can test one-bit slice of Figure 9.9a. This complete test set can be re-examined to identify the redundant vectors. These vectors can be deleted

ALU *Source Control*:

I2	I1	I0	R	S
0	0	0	0	A
0	0	1	0	T
0	1	0	D	A
0	1	1	D	T
1	0	0	A	A
1	0	1	A	T
1	1	0	D	0
1	1	1	A	0

ALU *Function Control*:

Function	I5	I4	I3	F output
f0	0	0	0	R + S
f1	0	0	1	S − R
f2	0	1	0	R − S
f3	0	1	1	R OR S
f4	1	0	0	R AND S
f5	1	0	1	$\bar{R}$ AND S
f6	1	1	0	R XOR S
f7	1	1	1	R XNOR S

ALU *Destination Control*:

I7	I6	Function
0	0	A ← L
0	1	T ← L
1	0	A,T ← L
1	1	NOP

Shift Control:

I9	I8	L output
0	0	F
0	1	R1
1	0	L1
1	1	Not used

Figure 9.9(b) Microinstruction control fields for Figure 9.9(a). (From [5] © 1981 IEEE. Reprinted with permission.)

from the final test set to achieve a minimal test set [6]. The overall test method can be summarized as follows:

Step 1: Identify individual functional blocks in one-bit slice.

Step 2: Calculate test sets for individual functional blocks. The tests are calculated at the functional level to verify the truth table of the block.

Step 3: Merge individual test sets to obtain a complete test set for a one-bit slice.

9.4.2 Testing of k-Bit Processor

By using a straight k-times replication, a k-bit processor can be made by using k copies of the one-bit slice of Figure 9.9(a). Such an expansion is shown in Figure 9.10. The test generation for this k-bit version is the same as the one-bit slice. For example, the tests for multiplexer blocks MR_i and MS_i are the same with D, A,

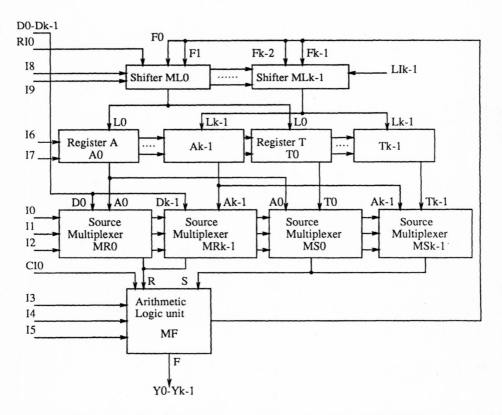

Figure 9.10 Extension to k-bit processor using one-bit slice. (From [5], © 1981 IEEE. Reprinted with permission.)

and T replaced by D_i, A_i, and T_i. Also, note that the testing of MR_i and MS_i can be done simultaneously with MR_j and MS_j, by 32 vectors, where $i \neq j$. Similarly, blocks MA_i and MT_i can be tested by 16 vectors, and ML_i can be tested by 32 vectors for all $i = 0$ to $(k - 1)$. The k-bit version of block MF has $(4 + 2k)$ inputs, and hence can be tested exhaustively by a sequence of 2^{4+2k} vectors. Thus, the total test length for the k-bit version is $80 + 2^{4+2k}$. If a general-purpose memory is also built with the k-bit processor, the test sequence for the memory block can be obtained by any method discussed in Chapter 7.

If a linear cascading of a one-bit slice is done to obtain k-bit operand, it can be tested by the test sequence of a one-bit slice. For example, a linear cascading of k copies of a one-bit slice is shown in Figure 9.11. The 114 vector long test

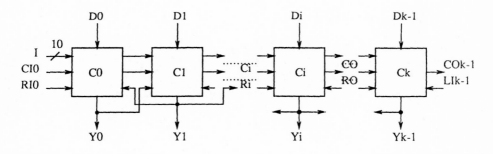

Figure 9.11 Extension to k-bit processor using linear cascading. (From [5], © 1981 IEEE. Reprinted with permission.)

sequence developed for a one-bit slice can be applied simultaneously to all the cells, and hence this k-bit version can be tested by 114 vectors. Because the test length remains constant for the k-bit version, it is popularly termed C-testability. Note that the structure of Figure 9.11 is analogous to the sequential circuit viewed at different time periods, as discussed in Section 8.4.2. The test vectors for Figure 9.11 also can be obtained by any method discussed in Section 8.4.2. If a k-bit bit-sliced system is not C-testable, it can be made so by using some extra hardware and external control [7].

Although this method is applicable to only bit-sliced systems, it has wide application due to the widespread use of such systems. The major advantage of this test method is that a small test sequence is calculated for one-bit slice, which is also applicable to a k-bit processor. The test sequence for a k-bit version is the same, except for the functional block, which is not obtained by linear cascading. Note that the length of test sequence is so small that it makes the test application time negligible.

9.5 CONCURRENT CHECKING

Microprocessors are the controlling units in a computer system. Apart from one-time testing before putting it in the field, verifying the fault-free operation of a microprocessor during its normal operation is important. Such testing is popularly known as *concurrent checking*, which not only detects permanent faults, but also transient errors. Many different methods have been developed for effective concurrent checking. Although a detailed discussion of these methods is beyond the scope of this book, a brief note on some major techniques is given below.

9.5.1 Error-Detecting Codes

A significant amount of work has been done using error-detecting codes for micro-processors to implement a concurrent error-detection capability. While much of the work is directed toward fault tolerance using duplicate circuits, self-testing schemes have also been developed. Various coding techniques such as parity codes, Hamming codes, *m*-out-of-*n* codes, Berger codes, two-rail codes, cyclic codes, self-checking checkers, *et cetera*, have been used. Coding theory is a broad area and beyond the scope of this book. A comprehensive discussion of different coding techniques is given in [8]. For microprocessors, concurrent checking techniques based upon error-detecting codes can be broadly divided into three categories:

1. Use of duplication circuits and output comparison checkers;
2. Extensive use of easily testable PLAs, as discussed in Chapter 6;
3. Use of coding methods for control, address, and data lines.

The implementation of any coding technique requires additional hardware, and hence we consider this topic as part of the design for testability. In general, self-testing of a complete system based upon a coding technique may cause about 30% to 60% hardware overhead. The exact implementation details on some widely used methods are given in Chapter 10. However, to illustrate the overall system, a self-testing processor organization using a coding technique is given in Figure 9.12. The detailed implementations of the control, ALU, and register sections are also given in Figure 9.12.

9.5.2 Check-Point Technique

The *check-point* method is generally used to detect two kinds of errors: (1) sequencing errors, when a processor executes instructions (or microinstructions) in the wrong sequence and (2) bit errors, when some bits in an instruction are in error. One method for testing is the check-point insertion in the program flow graph as described in [10]. In this method a *parallel signature analyzer* (PSA) or multi-input linear feedback shift register is used. A detailed discussion of the PSA is given in Chapter 10. During the normal execution of a microprogram, either a special field of microinstructions or all of the microinstructions are continuously monitored by the PSA. At predetermined check points, the state of the PSA or the signature is compared with the hard-wired reference signature. A mismatch indicates an error between the previous and present check point.

Two different methods have been suggested [10] to implement this method. One possibility is to use a special field (called *code field*) in every microinstruction. In this situation, only the code field is monitored by the PSA. The use of a code field is very attractive for horizontal microinstructions, where the width of the

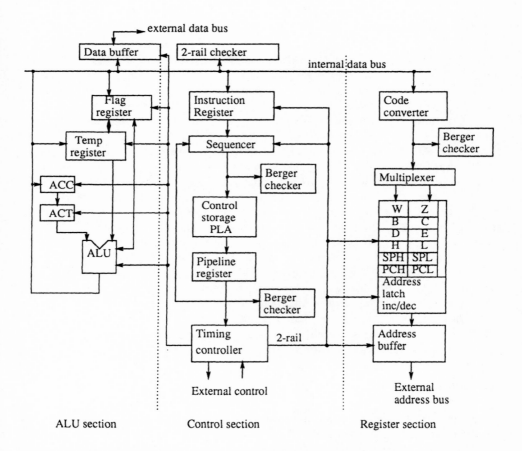

Figure 9.12 Architecture of a self-test microprocessor. (From [9], © 1988 IEEE. Reprinted with permission.)

microinstruction is large and the average number of microinstructions between two check points is small. However, this scheme only detects errors in the code field.

Another possibility is to use a special microinstruction for synchronization at the check points (called *synch instruction*). In this situation, the PSA monitors the entire set of microinstructions. This scheme is suitable for vertical microinstructions, where the width of the microinstruction is small and the average number of microinstructions between two check-points is large. This scheme can also be used for off-line testing of the control unit. In this scheme, the control monitor is large and the synch microinstruction requires additional ROM space. Also, there is a performance penalty due to the additional synch microinstructions.

9.5.3 Watchdog Processor

A watchdog processor is a co-processor, which is used to monitor the main microprocessor. This watchdog processor can be added to any system that supports a co-processor. A general survey of this method is given in [11], while a detailed control flow-checking scheme using a watchdog processor is given in [12].

Conceptually, this method is very simple. The information about the main microprocessor or the process under test is supplied to the watchdog processor (called the setup phase). Based upon this information, watchdog processor monitors the main microprocessor and concurrently collects the information about the memory access, control flow, or data manipulation. Error detection in the execution process of the main microprocessor is done by comparing the information collected by the watchdog processor and the reference information used during setup. Note that the technique based on PSA for concurrently checking the program flow, as discussed in the previous section, conceptually is a watchdog technique, where the PSA is equivalent to the watchdog processor.

The main disadvantage of this method is the requirement of a co-processor, which is significantly costly. Furthermore, if a system does not support a co-processor, it will require an architectural modification to implement this method. The main advantage of the method is that it can detect not only permanent faults, but also transient faults. Also, it eliminates concern about design modification or test generation for the main microprocessor.

9.6 SUMMARY

The complete testing of a microprocessor is extremely complex. To simplify this problem, functional fault models are used. The straightforward approach is to verify the complete instruction set. Machine instruction level verification and microinstruction level verification methods are given in this chapter. The microinstruction level verification method is comparatively easy and more comprehensive.

The testing method for bit-sliced processors has been discussed. Bit-sliced processors can be tested very efficiently by a small test sequence. This testing basically uses a divide-and-conquer rule. All the functional blocks in a one-bit slice are identified and their individual test sequences are calculated. These test sequences are merged into a complete test set. A k-bit processor obtained by linear cascading of one-bit slices can be tested by the same test set.

A brief discussion of concurrent checking methods is given. The basic idea of concurrent checking using error-detecting codes, check points, and a watchdog processor is discussed.

REFERENCES

1. C. Robach and G. Saucier, "Dynamic Testing of Control Units," *IEEE Trans. Comp.*, **27**(7), pp. 617–623, July 1978.

2. S.M. Thatte and J.A. Abraham, "Test Generation for Microprocessors," *IEEE Trans. Comp.*, **29**(6), pp. 429–441, June 1980.

3. D. Brahme and J.A. Abraham, "Functional Testing of Microprocessors," *IEEE Trans. Comp.*, **33**(6), pp. 475–485, June 1984.

4. G. Kruger, "Automatic Generation of Self-Test Programs—A New Feature of the MIMOLA Design System," *Proc. 23rd Design Auto. Conf.*, pp. 378–384, 1986.

5. T. Sridhar and J.P. Hayes, "A Functional Approach to Testing Bit-Sliced Microprocessors," *IEEE Trans. Comp.*, **30**(8), pp. 563–571, Aug. 1981.

6. S. Hwang, R. Rajsuman, and Y.K. Malaiya, "On the Testing of Microprogrammed Processor," *Proc. Int. Symp. Microprogramming and Microarchitecture*, pp. 260–266, 1990.

7. T. Sridhar and J.P. Hayes, "Design of Easily Testable Bit-Sliced Systems," *IEEE Trans. Comp.*, **30**(11), pp. 842–854, Nov. 1981.

8. T.R.N. Rao and E. Fujiwara, "Error-Control Coding for Computer Systems," Prentice Hall, Englewood Cliffs, NJ, 1989.

9. T. Nanya and T. Kawamura, "Error Secure/Propagating Concept and its Application to the Design of Strongly Fault-Secure Processors," *IEEE Trans. Comp.*, **37**(1), pp. 14–24, Jan. 1988.

10. T. Sridhar and S.M. Thatte, "Concurrent Checking of Program Flow in VLSI Processors," *Proc. Int. Test Conf.*, pp. 191–199, 1982.

11. A. Mahmood and E.J. McCluskey, "Concurrent Error Detection Using Watchdog Processors—A Survey," *IEEE Trans. Comp.*, **37**(2), pp. 160–174, Feb. 1988.

12. N.R. Sexena and E.J. McCluskey, "Control-Flow Checking Using Watchdog Assists and Extended-Precision Checksums," *Proc. Fault Tol. Comp.*, pp. 428–435, 1989.

Chapter 10
Design for Testability
by S. Gupta* and R. Rajsuman

10.1 INTRODUCTION

The notion of testability is very wide and includes the effort required for test generation, fault coverage analysis, the size of test set, test application time, and the cost of test equipment. In different implementations of a function, some are more testable than others. However, using a more testable implementation for one section of a chip may not simplify the overall testing of the complete chip. With increased complexity, the observability and controllability problem becomes worse. Difficulties in testing integrated circuits inspired the idea of design for testability. Design for testability is a large collection of different techniques. These techniques can be implemented in two ways: (1) the original design is made without considering testing, and then an engineer adds the test circuitry; (2) the design engineer considers testing from the beginning and incorporates test circuitry. Obviously, the second approach is more desirable.

A number of *ad hoc* techniques such as test point insertion and partitioning are widely used to simplify testing. In this chapter, more structured techniques are discussed to enhance testability. These techniques require adding extra circuitry into the original design and generally are considered from the beginning of a design. These methods may or may not require major changes in the design process. However, some minor restrictions are always expected to be considered by the design engineer.

10.2 SCAN DESIGN

The controllability and observability problem is severe in sequential circuits as discussed in Chapters 8 and 9. Efficient techniques to generate checking sequences

---

*S. Gupta is at the University of Southern California, Los Angeles.

for arbitrary circuits are not yet available. The difficulty in finding a checking sequence motivated the search for other methods for setting and observing the states. The SCAN design is one popular method to achieve this purpose. The basic philosophy behind the scan design is the easy access to some internal points in the circuit. These internal points are used as test points to apply the test vectors and to observe the circuit response.

10.2.1 Multiplexed Data Scan Design

Consider a generalized synchronous sequential circuit, as shown in Figure 10.1. The inputs to the combinational block are the primary inputs and present state lines from the flip-flops. The outputs of the combinational block are the primary outputs and next state lines. The flip-flops are assumed to be edge-triggered. Obviously, in this circuit, the present and next state lines are not directly observable and controllable. This circuit is modified by using some extra logic, as shown in Figure 10.2. An extra control input (test point P) is added. This extra control signal is used to switch the circuit into test mode (called *scan mode*, when $p = 1$). Further, a multiplexer, shown as SW in Figure 10.2, is added to the inputs of all the flip-flops. The logic level schematics of switch SW is shown in Figure 10.3.

This design is called *multiplexed data scan* (MD-scan) [1]. One data input from each multiplexer is connected to the next state signal, which was originally connected to the data input of a given flip-flop. Hence, if $p = 0$, the circuit performs exactly the same way as the original circuit. However, from the multiplexer, the other data input to the first flip-flop is now connected to one of the primary input pins. This input pin is called the *scan input*, since it is used to initialize the state during the scan mode ($p = 1$). The output of the first flip-flop is connected to the data input of the multiplexer connected to the next flip-flop. Hence, if $p = 1$, all flip-flops form a shift register. The circuit state can now be set to any desired value

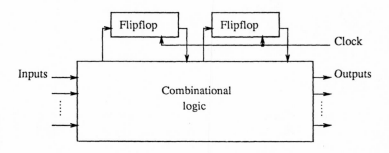

Figure 10.1 A generalized sequential circuit. (From [1], © 1973 IEEE. Reprinted with permission.)

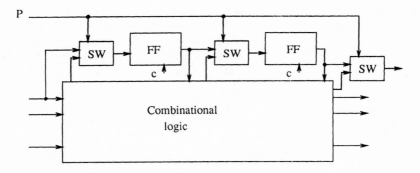

Figure 10.2 Modified sequential circuit. (From [1], © 1973 IEEE. Reprinted with permission.)

by serially applying the desired vector sequence at the scan-input and shifting it into the flip-flops. Also, the output of the last flip-flop is connected to an output pin through a multiplexer. Hence, when $p = 1$, not only a new state can be forced into the flip-flops, but also the old state value can be serially taken out and observed at the scan output pin.

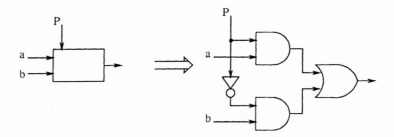

Figure 10.3 Detailed circuit of switch SW, two-bit multiplexer. (From [1], © 1973 IEEE. Reprinted with permission.)

Test generation for the sequential circuit is now greatly simplified. Any standard combinational test generation program such as PODEM or FAN can be used to generate a test vector for the combinational circuit. The scan chain is then used to set the present state lines to the value required by the test. Once the state is set, the rest of the inputs are applied at the primary input pins. The primary outputs are observed, and the next state value is captured in the scan chain. After this test application, the state can be observed by using the scan mode. Hence, the scan methodology obviates the need to derive initialization and distinguishing sequences, and reduces the complexity of sequential test generation into combinational test generation. The whole testing method is summarized as follows:

1. Switch the circuit into the scan mode, $p = 1$.
2. Test the scan chain by applying a sequence of alternating 1s and 0s. This ensures that the scan chain is capable of shifting in any data and that none of the flip-flop lines have stuck-at faults.
3. Serially shift in the state required by the next test vector. This is accomplished by setting the circuit into the scan mode ($p = 1$), and serially applying the next input to the scan input. Note that during this process, the primary inputs to the circuit can have any value.
4. Return to the normal mode, $p = 0$. Apply the primary inputs required by the test vector, and check the primary output values. Clock the flip-flops to capture the next state signals.
5. Switch to the scan mode and shift out (and check) the state of the circuit. Simultaneously, shift in the state required by the next test vector.
6. Go to step 4 if any more test vectors need to be applied.

In the above design, the storage elements are assumed to be edge-triggered flip-flops. If these are level-triggered, operating them as a shift register in the scan mode is not possible. Also, the flip-flops are modified by adding multiplexers at their data inputs. There are some optional ways to implement scanning, although the basic idea remains the same. The circuit has two modes of operation, one for normal operation and the second for controlling and observing the state. In the next section, a level-triggered latch-based scan design is given.

10.2.2 Level Sensitive Scan Design

The *level sensitive scan design* (LSSD) rules comprise latch design and clocking strategies [2–3]. These rules ensure that the steady state circuit operation is independent of the rise and fall time. The total delay through a given number of logic levels is bounded by a known value and depends only on the level of inputs, not on the exact order in which they change. This is accomplished by using a two-phase nonoverlapping clock. The basic storage element in this design is a two-flip-flop latch, as shown in Figure 10.4. The inputs to this flip-flop pair are:

D: *Normal data input*, which is connected to one of the next state output lines of the combinational logic;
C: *System clock Clk*1, used during normal circuit operation to latch the next state output of the circuit into the flip-flop L_1 via input D;

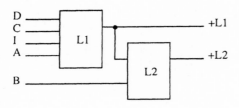

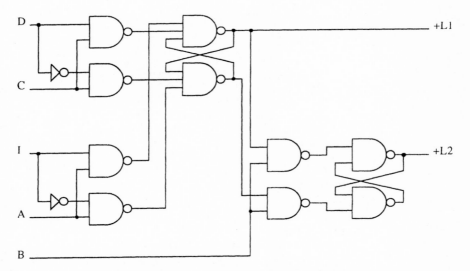

Figure 10.4 Schematics and detailed diagram of a polarity-hold SRL latch. (From [3], © 1977 IEEE. Reprinted with permission.)

I: *Scan data input*, connected either to the scan-in pin or to the output of L_2 of the previous flip-flop pair in the scan-chain;

A: *Scan clock*, which latches the scan data from I to the output of L_1.

B: *Clock Clk2*, which latches the output of L_1 to the output of L_2.

The LSSD structure is shown in Figure 10.5. In this design, the L_1-L_2 flip-flop pair latches either the serial scan data, or latches the next state output. Hence, this design differs from MD-scan, where, instead of an input data multiplexer, the use of a scan clock determines the mode of circuit operation. In Figure 10.5, the primary input and output pins are also marked. The next state outputs of the

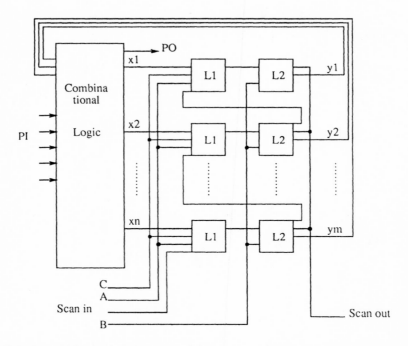

Figure 10.5 LSSD double-latch design. (From [3], © 1977 IEEE. Reprinted with permission.)

combinational circuit are x_1, x_2, . . . , x_n. These are connected to the data input D of L_1 flip-flops. The present state information is available at the output of L_2 flip-flops (Y_1, Y_2, . . . , Y_m). An application of Clk1 followed by the application of Clk2, latches the next state output of the circuit to the present state input.

The scan-chain is configured by connecting the I input of the first L_1 flip-flop to the scan-in pin. The inputs I of each of the other L_1 outputs are connected to the L_2 outputs of the previous flip-flop pair. Hence, a new state of the circuit can be scanned in by using clocks A-shift and B-shift, alternately. The main steps in the testing are summarized as follows:

1. Test the scan chain by applying a sequence of alternating 1s and 0s to the scan-in pin and using the clocks A-shift and B-shift.
2. Serially shift in the state required by the next test vector from the scan input using alternating A-shift and B-shift clocks.
3. Apply the primary inputs required by the test vector and check the primary outputs.
4. Use clocks A and B to clock in the next state output for the test vector into the flip-flop.

5. Serially scan out the state using clocks A-shift and B-shift, while scanning in the state required by the next test vector.
6. Go to step 3 if more test vectors need to be applied.

10.2.3 Pros and Cons

Although scan design greatly simplifies the testing of sequential circuits, there are some penalties in the use of it.

(1) Area Overhead

The modification of the storage element by adding multiplexers at the input of the flip-flops, or adding data and clock inputs to the L_1 flip-flop in LSSD, requires extra hardware. To form a shift register during scan mode, the output of each storage element must be routed to the input of the next flip-flop. In some circuits, the routing overhead may be large.

(2) Pin Count

Extra pins are required for the scan inputs. In MD-scan, one extra pin is needed to control the mode of operation. Here, the scan-in and scan-out can be multiplexed with normal circuit inputs and outputs. In LSSD, four extra pins are required for the A-shift, B-shift, data scan-in, and data scan-out.

(3) Performance

In the MD-scan design, two levels of extra delay are added between the next state output of the combinational circuit and storage elements. This delay is incurred during normal circuit operation as well. Hence, it increases the system clock period. This also decreases the circuit speed during normal operation. Note that this effect is minimized in LSSD because of the use of the extra clock, and an extra level of logic is not added. However, in both designs, due to the formation of a scan chain, the fan-out at the outputs of all storage elements is increased by one, thereby slowing the circuit.

(4) Test Time

Although scan design simplifies the testing process, the test time requirements may be higher than for a nonscanning circuit. Each test vector application in the scan design requires that the entire scan chain be serially loaded with the new state. Hence, the test time is a product of the length of the scan chain and the number of combinational test vectors. If the circuit has a long scan chain, the test time may be very high. This problem is remedied in a number of ways:

(a) Instead of configuring all the storage elements in the circuit in a single scan chain, these are configured into multiple scan chains. Each chain can have its own scan input and output. This reduces the test time because a number of state bits can be loaded in parallel. Multiplexing of input and output pins is done to minimize the pin overhead.

(b) Most combinational test vectors require the setting of only a few storage elements. The other storage elements already may be in the correct state, or required by the test vectors (don't-care). Similarly, for most of the test vectors, the fault effect may propagate to only a few storage elements. Hence, it is enough to set a small number of storage elements for each test. However, the serial nature of the scan chain requires a large number of clocks to accomplish even selective setting or checking of the state. Random access scan has been proposed to address this issue. In the scan mode, the storage elements are individually addressable for writing or reading. This avoids long shifts of the scan chain. Unfortunately, the area overhead in this scheme is significant, and hence it is not popular. In the next section, we describe partial scan design to address some of these issues.

10.3 PARTIAL SCAN

In the partial scan method, some of the flip-flops in the circuit are excluded from the scan chains. Figure 10.6(a) shows a sequential circuit with combinational blocks $C_1 \cdots C_5$ and storage elements $R_1 \cdots R_7$. In full scan, all the storage elements are connected in the scan chain, as shown in Figure 10.6(b). In partial scan, only some of the storage elements are connected in the scan chain, as shown in Figure 10.6(c), 10.6(d), and 10.6(e). In partial scan, there may be some flip-flops in the circuit that cannot be controlled or observed directly in the scan mode. In other words, if all the scan flip-flops are removed and their inputs and outputs are considered the primary inputs and outputs, the resulting circuit still is sequential. Partial scan provides a continuous range of designs between nonscanning and full-scan.

The main issue in partial scan is the choice of flip-flops to be included in the scan chain. One possible criterion is circuit performance that precludes the critical path from the scan chain. The other factors in the choice of flip-flops are fault coverage and the length of test sequence. In Ballast [4], storage elements are selected to be scanned, such that:

1. There are no cycles in the circuit.

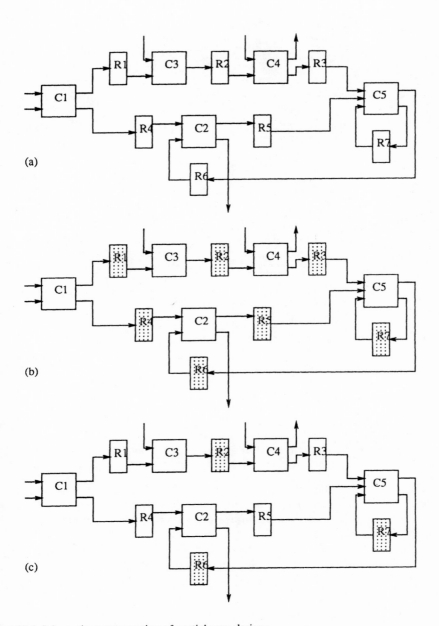

Figure 10.6 Schematic representation of partial scan designs.

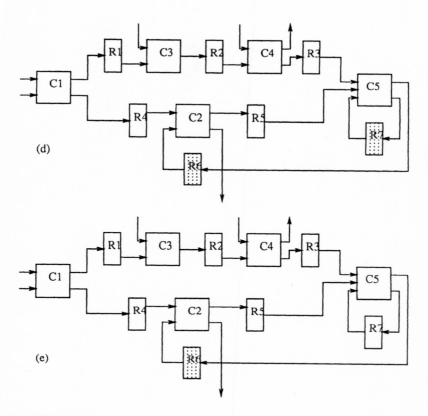

Figure 10.6 (Continued)

2. All paths between any two combinational blocks have the same number of unscanned storage elements such that these paths are balanced.

Tests for all faults in the resulting acyclic and balanced circuit (Figure 10.6(c)), can be generated by a combinational test generator. Another variation of this scheme is to scan storage elements that make the circuit acyclic, although different paths may not be balanced (Figure 10.6(d)). In this case, a sequential test generator is required, but the test generation complexity is reduced compared to nonscanning. In [5], the test generation complexity has been observed to be high in circuits that have long feedback cycles. In such cases, the number of flip-flops can be selected to break the feedback loops while excluding self-loops, as shown in Figure 10.6(e).

In general, a large number of storage elements meet the basic criteria to be connected in scan chain. A set with a minimum number of flip-flops may be selected to reduce the area overhead. However, flip-flops in the critical path should be left unscanned to limit the performance penalty.

One advantage of partial scan is that the area overhead is less compared to full-scan. In some cases, a small reduction in area may make the design acceptable. Partial scan also reduces the performance penalty compared to full-scan, which introduces extra delay at all next state lines. In general, only a few next state lines are in the critical path, leaving these flip-flops unscanned, and thus the performance penalty is minimized in partial scan. Also, full-scan needs additional modification if some part of the circuit is asynchronous. Note that this is not an inherent shortcoming of full-scan, but rather it is due to the fact that full-scan traditionally uses a combinational test generator. In partial scan, a sequential test generator is used, and hence asynchronous parts of the circuit can be handled without difficulty.

The recent development of viable sequential test generators has encouraged the use of partial scan. However, some partial scan designs still use combinational test generators. For some scan flip-flops, the fault coverage may be poor in partial scan.

10.4 BOUNDARY SCAN

Design-for-testability techniques, such as scan, can simplify the testing problem for the individual chip. Testing a computer system requires that all the chips on a board manufactured by different vendors have uniform scan implementation. Even if such is the case, the size of the board logic becomes too large for test-generation programs. Another problem is the testing of interconnection between chips. The failure modes are open and short circuits between adjacent pins or wires. Diagnosis of a failing board is necessary to replace the faulty chips.

In the past, the testing of boards has been done by probing through-the-hole devices. The probes were connected to the IC's pins for testing. Chip-level tests were used directly by driving the connections back to the other components. This technique is off-line and the board must be pulled off the system to connect the probes. Also, with new board designs, such as surface-mount technology, probing becomes impossible. Hence, a strong need to develop a comprehensive test strategy for board testing became the driving force behind the development of *boundary scan*. Since the boards contain chips from different manufacturers, the IEEE has developed a standard for the boundary scan (IEEE Standard 1149.1 [7], also known as JTAG).

10.4.1 Basic Concept

The basic concept behind boundary scan is to introduce a scan cell at each input-output (I/O) pin. The boundary scan cells are interconnected to form a scan chain.

The circuit of a boundary scan cell is shown in Figure 10.7(a). In addition to the boundary scan chain, the circuit can contain a scan chain to set or to observe the internal storage elements. These chains are called *internal scan chains*. Figure 10.7(b) shows a simple circuit with a boundary scan chain. This chain is used to shift in any desired test vector for the chip. The output response of the chip logic can also be shifted out by using the boundary scan chain. Hence, boundary scan allows for chip-level tests without the need to consider the board-level interconnections.

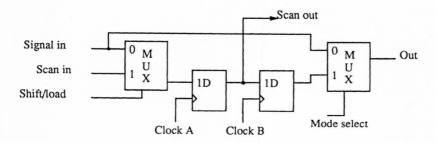

Figure 10.7(a) Schematic design of boundary scan cell.

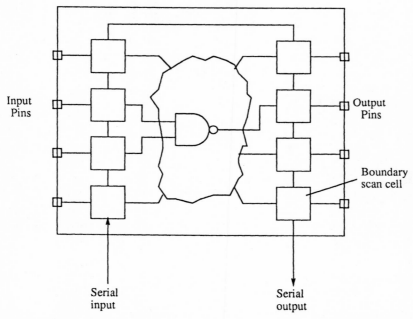

Figure 10.7(b) Boundary scan design to test on-chip logic. (From [7], © 1990 IEEE. Reprinted with permission.)

Because the pins of all the chips on a board can be individually set or observed, the interconnections between chips also can be easily tested. For example, a short circuit between two lines on the board can be tested by setting the pins driving the two lines to two different values. The observed values are captured by the pins being driven by these lines, and then these values are shifted out. Figure 10.8 shows the concept. The main advantage of using boundary scan for interconnect lines testing is that the tests can be derived solely on the interconnection topology and independent of the nature of the chips.

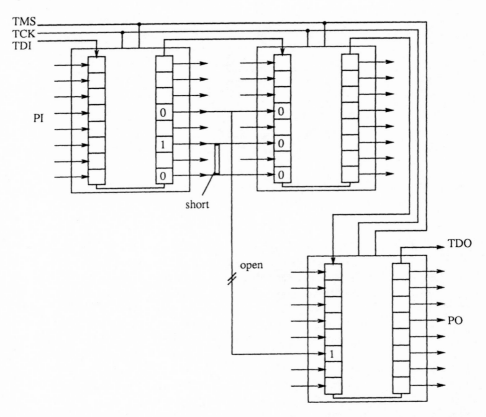

Figure 10.8 Testing interconnect using boundary scan. (From [7], © 1990 IEEE. Reprinted with permission.)

10.4.2 Test Access Port

The architecture of the IEEE 1149.1 boundary scan standard is shown in Figure 10.9. This consists of a four-bit *test access port* (TAP), which includes a test clock,

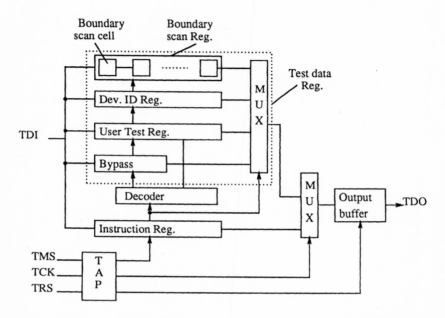

Figure 10.9 Architecture of IEEE 1149.1 standard. (From [6], © 1989 IEEE. Reprinted with permission.)

TCK, test mode select signal, TMS, test data input, TDI, and test data output, TDO. An instruction is serially shifted into the instruction register via the TDI pin. Decoding logic is used to decode the instruction and to generate the appropriate control signals. In Figure 10.9, the boundary scan chain is shown as the boundary scan register.

The TAP inputs and outputs may be connected at the board level. Figure 10.10 shows three different interconnection configurations. In each configuration, the test bus may be controlled by an ATE or a bus master, which controls the test bus and provides the interface at the board backplane. In Figure 10.10(a), the TMS and TCK signals are fed from the bus master to form a boundary scan chain. In Figure 10.10(b), a pair of TMS signals (TMS1 and TMS2) ensure that only one serial path is scanning at a given time. Figure 10.10(c) shows four serial paths that have separate TDI and TDO signals, but are controlled by the same TMS and TCK signals.

Because the boundary scan registers of all chips on the board are serially chained, testing time can be significant. The testing is not time-efficient, especially when some chips are not being tested. A one-bit bypass register is provided to eliminate delays through long boundary scan chains of untested chips. The user

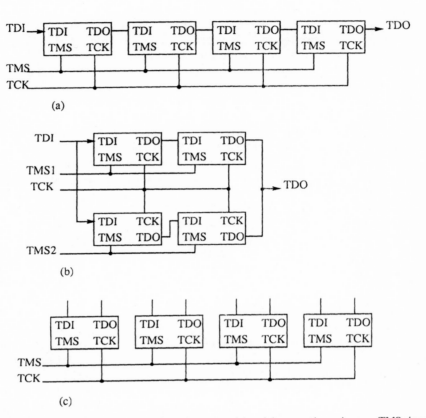

Figure 10.10 Various configurations during boundary scan: (a) serial connection using one TMS signal; (b) connection in two parallel serial chains; (c) multiple independent paths with common TMS and TCK signals. (From [7], © 1990 IEEE. Reprinted with permission.)

test register may be the internal registers, which may also be serially loaded via the TDI pin and observed at the TDO pin. Device ID register is optional. The multiplexer is used for selectively shifting out the contents of one of the registers. To test this board, first, the TMS line is used to set the mode of the TAP controller. Then the TDI line is used for serially shifting in the next instruction. The instruction is decoded and TDI may be used to scan in the data for the boundary scan, user, or bypass register. The test is applied and the circuit output is captured in the scan registers. The captured data are then serially shifted out via the TDO pin. The various test modes supported under this architecture are the following:

1. *Extest:* This instruction allows the testing of the board-level interconnection. The instruction is loaded via the TDI pin. After the boundary scan register

is decoded, it is loaded with the test data via TDI. The test is then applied and the response is captured in the boundary scan register. The contents of the boundary scan register are then shifted out.

2. *Sample/Preload:* This mode is used either to capture the chip output during normal circuit operation or to preload a value into the boundary scan register. The data at the input-output pins can be captured in the boundary scan chain without interrupting normal circuit operation. The data can then be shifted out for examination. Similarly, the test data can be preloaded into the boundary scan chain without interrupting normal circuit operation.

3. *Intest:* This mode is used to test the chip's internal circuitry. This is accomplished by shifting in the test vector into the boundary scan chain, applying the test to the internal logic, and capturing and shifting out the response by using boundary scan chain. Note that this testing may be very slow because the test data are shifted in and out serially.

4. *Runbist:* This instruction provides the capability of using BIST circuits to test the internal circuits. However, IEEE Standard 1149.1 does not specify any procedure or implementation of BIST methods.

10.5 CROSS-CHECK DESIGN

Recently, a new methodology (called *cross-check*) has been developed to overcome the problem of poor observability and controllability [8]. In this technique, all primitive logic elements are designed with an extra pass transistor at their outputs. This transistor behaves as a test point. A circuit made of these modified cells is placed and routed into an orthogonal grid made up of probe lines (*P*) and sense lines (*S*). The gate of the pass transistor is controlled by *P* (an external control), and the value obtained through it is observed by *S* lines. Interestingly, this concept is similar to that shown in Figure 4.7. Effectively, it looks like the circuit has been placed on a bed of nails, such that each nail, behaving as a probe, senses the output of a gate. This concept is shown in Figure 10.11(a). A test controller controls a four-pin bus to activate the probe lines and receive data through the sense lines. The observed data are fed to an MISR for compression and to form a signature. The overall structure is shown in Figure 10.11(b).

Note that the same grid of probe and sense lines also can be used to control any internal node in the circuit. If there are memory elements in the circuit, they can be forced to a desired value by using this grid, as shown in Figure 10.11(c).

The main advantage of the cross-check method is that it provides 100% observability. If a circuit has been designed according to scan rules (or LSSD), it already has four extra pins dedicated to testing. These four pins also are used to control the probe and sense lines, and hence this method does not require additional pins. The only disadvantage of this method is that the extra transistor at the output

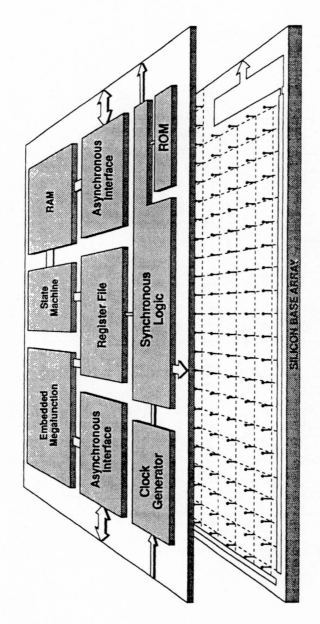

Figure 10.11(a) ASIC design and base array; bed of nails concept of the cross-check method. (Courtesy of CrossCheck Technology, Inc.)

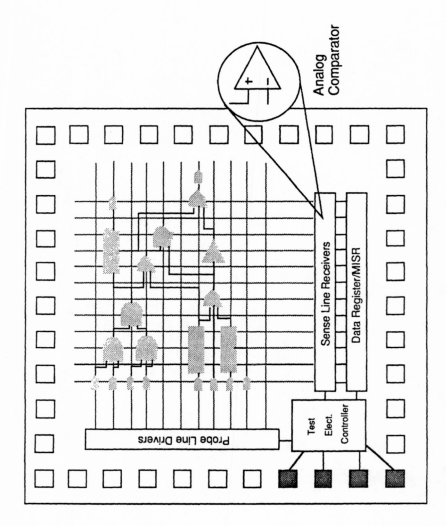

Figure 10.11(b) CrossCheck ASIC; overall structure of the cross-check method. (Courtesy of CrossCheck Technology, Inc.)

215

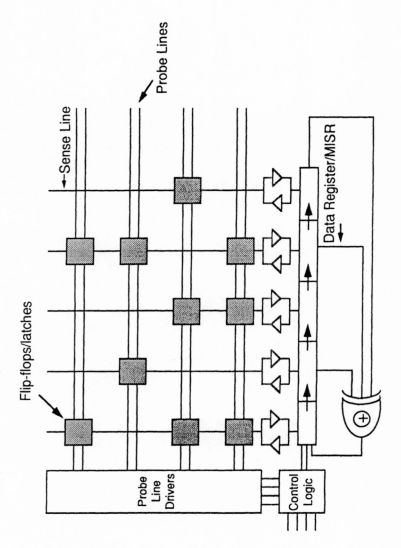

Figure 10.11(c) CrossCheck controllability in the base array; controllability of internal flip-flops using a grid. (Courtesy of CrossCheck Technology, Inc.)

of each gate and the routing of probe and sense lines results in large area overhead. This area is an additional penalty beyond the overhead due to scan implementation. Also, the extra transistor at the output of every gate increases the load capacitance, and hence causes a speed penalty.

10.6   BUILT-IN SELF-TEST

Traditionally, circuits are tested by *automatic test equipment* (ATE), which can store the test vectors and expected response. During testing, the ATE output ports are connected to the inputs and outputs of the circuit (including scan inputs, clock, et cetera). The drawbacks of this approach are that the test equipment is very expensive and the test application rate is slow due to the long off-chip connections to and from the ATE. Also, testing with an ATE is not completely automatic because the ATE probes need to be manually connected to the circuit's I/O.

These shortcomings motivated the development of built-in self-test (BIST) methods. In BIST, extra circuitry is added around the original circuit to test the original circuit. Most of the disadvantages of external test equipment are eliminated by BIST. Two important components required in BIST are (1) *on-chip test pattern generator* (TPG) and (2) *on-chip test response compressor*. Conceptually, BIST can be viewed as in Figure 10.12. Generally, the TPG is implemented to generate random or pseudorandom test vectors. The volume of test response is large, and directly storing it on-chip is not practical. Hence, the test response is compressed. At the end of the testing session, the compressed value of the test response is compared with the expected fault-free value.

BIST can be implemented in two ways, by explicit and implicit testing. In explicit testing, a fixed test sequence is applied to the *circuit under test* (CUT) and the test response (or its compressed value) is compared with that of the fault-free circuit. Due to its nature, explicit testing requires the CUT to suspend normal operation and enter into a *test mode*. Implicit testing requires multiple copies of the circuit. During normal operation, the output of these copies is compared with each other. In this case, the two copies of the circuit are implicitly tested without using a test vector.

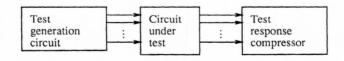

Figure 10.12 General structure of BIST.

Here we describe only explicit testing methods. The major concerns in an explicit BIST method are the nature of the test vector generator, the test response compressor, fault coverage, hardware overhead, and small testing time. Also, the extra hardware should be easy to design and fault coverage easy to determine. In the next two sections, we will discuss these factors in detail.

10.7 TEST PATTERN GENERATORS

In BIST, the length of the test sequence is not critical because testing is done at circuit speed. However, due to the fact that the entire test sequence has to be stored or generated on-chip, the test sequence must be either very short or easy to generate. Because the length of the test sequence affects the response compression process, test generation should also take into account the nature of the test compressor.

10.7.1 Deterministic Test Pattern Generators

Deterministic test generation methods for combinational circuits were discussed in Chapter 5. Major techniques for sequential circuits were discussed in Chapter 8. These techniques can generate test vectors that can be stored in an on-chip ROM. During self-test, these vectors can be applied to the CUT in a predetermined order. Because the hardware overhead or size of the ROM is directly dependent on the length of the test sequence, a short test sequence is desirable. One advantage of this scheme is that the vectors can be ordered as desired, and thus a better response compression can be obtained. Most test pattern generation programs perform fault simulation on the circuit with a short random sequence to eliminate easy to detect faults. The algorithmic test generator is then used to generate tests for the remaining faults. This reduces the overall test generation effort without drastically increasing the length of the test sequence.

Instead of using ROM to store deterministic test patterns, a finite state machine can be used to generate the required test sequence. Unfortunately, the size of the TPG is quite large in both approaches, and hence this scheme is rarely used.

10.7.2 Pseudorandom Test Vectors

Pseudorandom test patterns are widely used for testing digital ICs. These vectors can be generated by simple circuits. The main concern in this approach is to choose the right test generator that provides a high fault coverage, while keeping the test sequence length short. For a given pseudorandom sequence, fault coverage is generally obtained by simulation. If the fault coverage is found unacceptable, the length of test sequence is increased.

For large circuits, when simulation is expensive, probabilistic measures are used to compute the length of the test sequence for the desired fault coverage [9–10]. The length of test sequence for desired fault coverage should be much shorter than for an exhaustive test sequence, i.e., $N \ll 2^n$, where N is the number of test vectors in the pseudorandom test sequence and n is the number of primary inputs in the CUT. If such is not the case, *pseudoexhaustive testing* can be used because it does not require a separate fault coverage analysis. Below, we describe the two most commonly used random pattern generators.

10.7.2.1 Autonomous Linear Feedback Shift Register

Autonomous linear feedback shift registers (ALFSRs) are the most common random pattern generators used in BIST. The general structure of an ALFSR is shown in Figure 10.13(a). ALFSR comprises of an n-bit shift register to which feedback connections are added, using XOR gates. The output of the last stage of the shift register is selectively fed back to the inputs of the previous stages. The ALFSR is characterized by its feedback polynomial. The feedback polynomial for an n-degree ALFSR is given as

$$\phi(x) = \phi_n x^n + \phi_{n-1} x^{n-1} + \cdots + \phi_1 x + \phi_0 \tag{10.1}$$

where $\phi_i = 1$ indicates the presence of feedback in the input of D_i.

Example 1. Consider the three-stage ALFSR shown in Figure 10.13(b). The output of flip-flop D_2 is fed back to the inputs of D_0 and D_1. The feedback polynomial for this ALFSR is given by

$$\phi(x) = x^3 + x + 1 \tag{10.2}$$

Note that x^3 signifies that the output of D_2 is the feedback signal, x, and the one

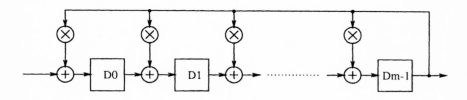

Figure 10.13(a) Autonomous linear feedback shift register.

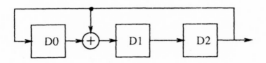

Figure 10.13(b) ALFSR for Example 1.

terms are present because the output is being fed back to the inputs of D_1 and D_0. Note that the x^2 term is not present because there is no feedback to the input of D_2. If the shift register is initialized to zero, the application of a clock does not change the state and the output of the shift register is zero. Hence, an all-zero sequence is produced at the output. If a nonzero initial value is used, a nonzero sequence appears at the output. For example, if D_0 is initialized to 1, and all other flip-flops are 0, the sequence generated at the output is: [0010111001011100101111. . .]. Note that the sequence repeats itself after every seven clock cycles.

The maximum period of a sequence generated by an n-stage ALFSR is $2^n - 1$. The output sequence is a function of the state of the ALFSR alone because it has no other input signal. Also, if the ALFSR is in the zero state, it stays in that state. Thus, only the state transitions that do not include the all-zero state need to be considered. The feedback polynomial determines the period of the sequence generated by the ALFSR. The feedback polynomials that generate maximum length sequences are called *primitive polynomials*.

The sequence generated by the ALFSR has certain properties:

1. The number of 1s and 0s in the sequence differs by 1.
2. There are an equal number of runs of 1s and 0s. The number of runs of length 1 occurs twice as often as the number of runs of length 2, which occurs twice as often as runs of length of 3, and so on.
3. Any maximal length sequence of length $2^n - 1$, and the sequence obtained by its nonzero cyclic shift differs in exactly $2^{n-1} - 1$ places. This means that the autocorrelation is high only when the shift value is zero.

When the ALFSR is used to generate pseudorandom patterns, the outputs of all the flip-flops in the shift register are used, instead of a single output of D_{n-1}. Thus, at each clock cycle, the n-bit state of the ALFSR is used as a test vector. As stated earlier, the ALFSR with primitive feedback polynomial cycles through all possible $2^n - 1$ nonzero states. Hence, irrespective of the initial condition, all the nonzero n-bit vectors are generated in the test sequence. This is useful because it does not preclude any test vector from occurring in the generated pseudorandom sequence.

10.7.2.2 Cellular Automata

A one-dimensional *cellular automata* (CA) is formed by interconnection of a one-dimensional array of flip-flops. The input to each stage of the CA is a function of its own and its two neighbors' outputs [11]. In BIST applications, in general, only linear functions are used. Figure 10.14(a) shows an example of CA [14]. Because the next state of stage i can only be dependent on the current state values of the $(i - 1)$th, ith, and $(i + 1)$th stages, it is a three-input and one-output function. The states of the three stages can take the eight possible values from 000 to 111. The rules by which a cell determines its next state are shown in Figure 10.14(b), where the first row represents the decimal number of the cell. The second row shows eight triplets, representing the current values of the $(i - 1)$th, ith, and $(i + 1)$th cells. The next state of the cell is written below the current value triplets. This group of eight-bits can take a value from 0 to 255, called the *rule number*. Figure 10.14(b) shows most commonly used rule numbers, 90 and 150. The bottom row in Figure 10.14(b) shows weights associated with the corresponding input combination. For a given rule, the decimal weights of all the inputs for which the functional output is 1 are added to obtain its rule number. Note that, out of 256 possible rules, only a small fraction represent linear functions.

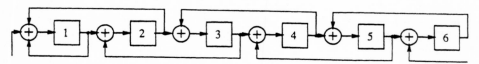

Figure 10.14(a) Example cellular automata. (From [14], © 1990 IEEE. Reprinted with permission.)

The nature of interconnection affects the quality of the sequence generated by a CA, which can be designed to generate maximal length sequences by using an appropriate combination of rules to interconnect the stages. Certain combinations of rules 90 and 150 can generate these sequences. Note that in a six-stage CA, as shown in Figure 10.14(a), the input to stage 1 is a function of the output of stages 1 and 2, and the third input is grounded. There are two possibilities at the boundary of a one-dimensional CA. One possibility is to assume that stage 6 is the left neighbor of stage 1. This is called the *circular boundary condition*. A second possibility is a *null boundary condition*, as shown in Figure 10.14(a). Studies of CA have shown that the boundary conditions are not critical to the nature of the patterns generated. Hence, to avoid end-to-end connections, null boundary conditions are generally used.

Because only nearest neighbor connections are needed, CAs are easy to build (especially for large n). This also makes them more modular and extensible than ALFSRs. A comprehensive comparison of patterns generated by LFSRs and CAs is given in [11]. The number of statistical tests are used to determine the randomness of a sequence. The sequences generated by CA have been found to pass more statistical tests than the sequences generated by the LFSR. This is attributed to the fact that the sequence generated at D_{i+1} in LFSRs might be obtained by shifting the sequence generated at D_i. This leads to some correlation between the vectors generated on consecutive clock cycles. In CA, due to the richness of interconnection, the correlation is lower. However, the number of XOR gates required in CA is usually large compared to the ALFSR.

	7	6	5	4	3	2	1	0
	111	110	101	100	011	010	001	000
Rule 90	0	1	0	1	1	0	1	0
Rule 150	1	0	0	1	0	1	1	0
	128	64	32	16	8	4	2	1

Figure 10.14(b) Naming of CA rules. (From [14], © 1990 IEEE. Reprinted with permission.)

10.7.2.3 Weighted Random Patterns

This approach addresses the problem by biasing the random pattern generator to generate the required test patterns with higher probability. Typically, after the

application of a short pseudorandom sequence, fault simulation is done and all detected faults are dropped. Afterward, a test generator is used to generate test patterns for some (or all) of the remaining faults. Generally, some of these faults require a specific pattern at the primary inputs. Note that the pseudorandom test generators generate 0s and 1s with equal probability. Once the weights are obtained for all primary inputs, a weighted random pattern circuit is built by using ALFSR [12], where it is modified with extra circuit to increase or decrease the probability of occurrence of 1 at specific inputs.

10.7.2.4 Test Point Insertion

In this approach, hard to detect faults are identified and the circuit is modified by inserting a test point near these fault sites. This increases the controllability or observability of hard to detect fault sites. Basically, these extra inputs provide access to the internal points in the circuit.

10.7.3 Pseudoexhaustive

If the circuit under test has a small number of inputs, it can be tested for all detectable faults by applying all possible 2^n test vectors. The test sequence can be easily generated by using ALFSRs or CAs, discussed above. Also, any binary counter can be used to generate an exhaustive test sequence. However, if the circuit has a large number of primary inputs, this scheme is not practical because test time can be extremely high. In this case, pseudoexhaustive testing techniques can be used.

10.7.3.1   Partial Dependent Circuits

Partial dependent (PD) circuits are multiple output circuits in which none of the outputs depends on all the primary inputs. A large number of circuits are PD. In this case, each output is exhaustively tested (separately or concurrently) with much less than 2^n patterns [13]. Note that, in this case, all single and multiple faults are detected. The following example illustrates pseudoexhaustive testing.

Example 2. Consider the circuit shown in Figure 10.15(a). This is a three-input and 2-output circuit. The output f is a function of the inputs w and x only. Similarly, output g depends on inputs x and y only. Because neither of the two functions depend on both w and y, one may connect w and y together and apply all four vectors as shown in Figure 10.15(a). Now $f(w, x)$ is exhaustively tested, as is $g(x, y)$. Hence, in this case, pseudoexhaustive testing has reduced the number of tests from eight (exhaustive) to four.

If we consider the circuit shown in Figure 10.15(b), each output depends on two inputs. In this circuit, there is no pair of inputs that does not appear together at any output. Hence, the above scheme cannot be used. However, four test vectors are sufficient to test the circuit shown in Figure 10.15(b). One can easily verify that all these vectors exhaustively test each function. Note that each nonzero test vector has a constant weight (number of 1s) of 2. Also, for each of the four vectors, $y = w$ XOR x. One can show that constant weight vectors or linear sums can be used to reduce the number of test vectors required to test any PD circuit.

10.7.3.2   Complete Dependence Circuits

Complete dependence (CD) circuits have at least one output that is a function of all the primary inputs. In this case, partitioning the circuit is possible to reduce the number of inputs on which an output depends. This is achieved by adding multiplexers, which are used to partition the circuit during the test mode. This can

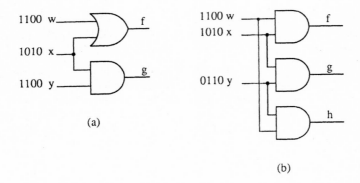

Figure 10.15 Example of pseudoexhaustive testing. (From [13], © 1984 IEEE. Reprinted with permission.)

reduce the pseudoexhaustive test sequence by converting the circuit to partial dependence during the test mode. This partitioning also is useful for PD circuits, which have outputs that are functions of a large number of inputs. This method requires additional hardware to be added to the circuit, and thus increases the delay in the circuit.

Another option is to use sensitized partitioning. In this scheme, some inputs are fixed to a value to sensitize certain parts of the circuit that partition the network. These partitions then can be tested individually. The advantage of this method is that it does not require extra hardware to partition and does not insert additional delay in the circuit during normal operation.

10.8 RESPONSE COMPRESSION FOR BIST

The amount of data in test response of a circuit during self-test is $xN \cdot m$, where x is the number of inputs, N is the number of test vectors, and m is the number of primary outputs. This large volume of data needs to be compared with the response of a fault-free circuit. To store this output response on-chip and make a bit-by-bit comparison is not practical. Hence, in all BIST methods, the test response is compressed. At the end of a test session, this compressed value is used as a signature and compared with the signature of a fault-free circuit.

The main problem with compression schemes is that some information is lost. Hence, even if the signature of CUT is the same as a fault-free signature, there are three possibilities:

1. The circuit under test is fault-free.
2. The CUT is faulty, but the fault has not been sensitized by a test pattern used during self-test.

3. The CUT is faulty, the fault has been sensitized, and it has been detected by one or more vectors, but the information has been lost during compression. Thus, the final signature is the same as a fault-free signature.

If the length of a test sequence is large, the probability that a fault will not be sensitized and detected is low. However, the loss of information during compression still may lead to the wrong conclusion. This phenomenon is called *aliasing*, and is used as a metric (along with hardware overhead) to evaluate the compression technique.

10.8.1 Parity Testing

This is the simplest compression technique. Consider a single output circuit with a fault-free response of $R = r_0 r_1 \cdots r_{N-1}$. The parity of the sequence is given by

$$S = r_0 \text{ XOR } r_1 \text{ XOR } \cdots \text{ XOR } r_{N-1} \tag{10.3}$$

One can easily see that the circuit shown in Figure 10.16 performs parity compression. Note that the circuit in the figure is equivalent to a single-stage LFSR, the input of which is connected to the output of the circuit under test. The feedback polynomial for this LFSR is

$$\phi(x) = x + 1 \tag{10.4}$$

The hardware overhead in this scheme is low and the final response is easy to check. If a single output circuit is exhaustively tested and the parity is odd, all single primary input stuck-at faults are detected by parity compression [15]. A fault in the circuit that causes an odd number of errors at the output is detected, but all faults that cause an even number of errors at the output are not detected. Hence, the probability of aliasing in this scheme is large. However, it can be augmented to compute more than one parity bit. In [15], a parity compression scheme is presented, where the output parity is augmented by the parity of the logical AND of the ith input with circuit output for all $0 \le i \le (n - 1)$. The aliasing probability decreases at the expense of higher compression complexity.

10.8.2 One-Count Testing

The one count compression counts the number of 1s in the test response. Hence, the *test response comparator* (TRC) is just a counter. The fault-free signature in this scheme is given by

$$S = \sum_{i=0}^{N-1} r_i \qquad (10.5)$$

where N is the number of test vectors. The number of bits required in this counter is $\leq \lceil \log_2 N \rceil$. This implies that the size of the compressor increases with increased length of the test sequence and is independent of the size of circuit. Any fault that causes an odd number of errors at the output is detected by this method. This is because an odd number of errors causes an unequal number of 1-to-0 and 0-to-1

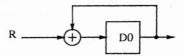

R

D0

Figure 10.16 Parity compressor.

changes, thereby changing the final count. The number of N-bit sequences that have the same 1s count as the fault-free circuit is $\binom{N}{S}$. There are 2^N possible output responses. If we assume that, in the presence of a fault, the CUT output will transform into one of the $2^N - 1$ possible erroneous N-bit sequences with equal probability, the aliasing probability is given by

$$P_{aliasing} = \frac{\binom{N}{S}}{2^N - 1} \qquad (10.6)$$

The numerator of the expression is a function of S. For low and high values of S, the numerator is small, but as S approaches $N/2$, the numerator increases, causing high aliasing. Hence, if the ones count compression is used, the TPG should be designed to have either a small or large S.

10.8.3 Syndrome Testing

Syndrome testing can be viewed as a special case of one count. The syndrome of a single output function is defined as the 1s count when the circuit is exhaustively tested, divided by the number of test vectors. Hence, the syndrome is the number of *minterms* in the function, divided by the size of input space.

A function f is called unate in its ith input x_i, if x_i or $\overline{x}_i$ appears in the expression f. Any s-a-1 or s-a-0 fault in a unate function can either add a new minterm in the function (leaving current minterms unchanged), or delete some existing minterms (without adding a new minterm). Unate functions are a small part of all possible functions and are syndrome testable. The faults that cause some minterms to be added and an equal number of original minterms to be deleted from the function are detectable. The faults that cause some minterms to be added and a different number of minterms to be deleted also are syndrome testable. By using extra circuitry and external control lines, any circuit can be modified to make it syndrome testable [10].

10.8.4 Transition Count

In transition count, the number of 1-to-0 and 0-to-1 transitions are counted in the test response. Schematically, this is shown in Figure 10.17. Transition count for a given sequence R is given by

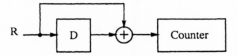

Figure 10.17 Transition count compression.

$$S = \sum_{i=0}^{N-2} (r_i + r_{i+1}) \qquad (10.7)$$

Here, Σ denotes decimal addition and $+$ represents modulo-2 addition. Note that in Figure 10.17, the D flip-flop needs to be initialized to r_0 to compute the transition count. The number of output sequences having same transition count is $2\binom{N-1}{S}$. Hence, if in the presence of some fault, all possible 2^N outputs are equally likely, the probability of aliasing for transition count is

$$P_{aliasing} = \frac{2\binom{N-1}{S}}{2^N - 1} \qquad (10.8)$$

If the fault-free transition count is either low or high, the probability of error escape is low. If the test vectors are applied in a specific order, zero aliasing can be achieved for transition count. Let T^0 (or T^1) be the set of test vectors that produces a 0 (or 1) at the output of a fault-free circuit. If a test sequence is obtained by alternating test vectors T^0 and T^1, the circuit is transition count testable.

10.8.5   Signature Analysis

Signature analysis is the most widely used compression scheme for BIST. In this scheme, the output of the circuit is connected to the input of the least significant bit of LFSR, as shown in Figure 10.18. The LFSR is the same as the ALFSR used for random pattern generation. This LFSR is initialized to all-zero state before test response compression. Consider the feedback polynomial for the n-degree LFSR as

$$\phi(x) = \phi_n x^n + \phi_{n-1} x^{n-1} + \cdots + \phi_1 x + \phi_0 \tag{10.9}$$

Let $R(x) = r_0 r_1 \cdots r_{N-1}$ be the output of the circuit, where r_0 is the response to the first vector, and r_{N-1} is the response to the last vector. If the circuit output $R(x)$ is compressed by using an LFSR with feedback polynomial $\phi(x)$, its signature $S(x)$ is given by

$$R(x) = Q(x)\phi(x) + S(x) \tag{10.10}$$

The final content of the LFSR, $S(x)$, is the remainder of the division of the circuit response $R(x)$ by the feedback polynomial $\phi(x)$. Note that the degree of the signature $S(x)$ is always less than the degree of feedback polynomial. The conditions under which the erroneous response will have the fault-free signature can now be easily characterized. Let the faulty circuit response be $R^f(x)$, where $R^f(x) \neq R(x)$. The signature for this response is

$$R^f(x) = Q^f(x) \cdot \phi(x) + S^f(x) \tag{10.11}$$

For aliasing, the faulty signature should be identical to the fault-free signature, although the faulty response is different from the fault-free response. Hence, $S^f(x) = S(x)$, while $R^f(x) \neq R(x)$. Taking the linear sum of the Equations (10.10) and (10.11), we obtain

$$R(x) + R^f(x) = Q(x) \cdot \phi(x) + S(x) + Q^f(x) \cdot \phi(x) + S^f(x) \tag{10.12}$$

Because $S^f(x) = S(x)$, this reduces to

$$R(x) + R^f(x) = \phi(x) \cdot [Q(x) + Q^f(x)] \tag{10.13}$$

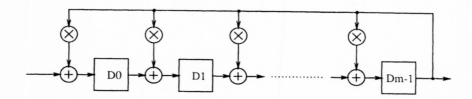

Figure 10.18 An m-bit signature analyzer.

Let the difference between fault-free response and the faulty response be $E(x) = R(x) + R'(x)$. Because all operations are modulo-2, the conditions under which an error $E(x)$ will cause aliasing is given by

$$E(x) = \phi(x) \cdot [Q(x) + Q'(x)] \tag{10.14}$$

Two important points should be noted:

1. Aliasing occurs when the error polynomial is a multiple of the feedback polynomial. Thus, the error polynomial causes aliasing if and only if it is divisible by the feedback polynomial of the LFSR.
2. Aliasing does not depend on the nature of fault-free response $R(x)$, but only depends on the error. This is different from the ones count and transition count compression, where the nature of the fault-free response directly influences aliasing.

The aliasing probability can be computed under the assumption that all 2^N possible error polynomials are equally likely. From the above discussion, we know that all the error polynomials which are multiples of $\phi(x)$ cause aliasing. Let the degree of $\phi(x)$ be m and the degree of $E(x) \leq (N - 1)$. The given $\phi(x)$ can be multipled with all possible 2^{N-m} polynomials of degree $\leq (N - 1 - m)$ to obtain all error polynomials which are multiples of $\phi(x)$. Note that this includes the error-free case, when $E(x) = 0$. Hence, there are $2^{N-m} - 1$ nonzero error polynomials that cause aliasing. Because all 2^N error polynomials are assumed to be equally likely, the aliasing probability is given by

$$P_{aliasing} = \frac{2^{N-m} - 1}{2^N - 1} \tag{10.15}$$

For the case, when $2^N \gg 1$, and $2^{N-m} \gg 1$, the above result reduces to

$$P_{aliasing} = 2^{-m} \tag{10.16}$$

Thus, the aliasing probability decreases exponentially with the increased length of LFSR. This equation also raises questions about the validity of the error model. The aliasing probability derived above is independent of the nature of the feedback polynomial. It is valid even if $\phi(x) = x^m$, when the output is not connected to any feedback taps. This reduces the LFSR to a simple shift register. However, the above analysis still yields the same aliasing probability. As, in this case, the signature comprises only the last m bits of the test response, it indicates that, if the equiprobable error model were true, there would be no need to apply a test sequence longer than m. This clearly indicates that the error model is not accurate. In the recent literature, errors in the test response are assumed to be independent. Thus, any bit r_i in the response may be in error with equal probability p. Also, an error in any bit is assumed to be independent of an error in any other bits. This model reduces to the previous case, when $p = 0.5$. Further discussion of error models is given in Appendix A.

The question of selecting the appropriate feedback polynomial for the LFSRs has been a subject of wide interest. In [16–17], the aliasing under an independent error model is shown to be less if the test response is compressed by using LFSRs with a primitive polynomial. The computation of aliasing probability for non-asymptotic test lengths has been widely studied. The exact expression for aliasing probability is given as [18]:

$$P_{aliasing}(p) = 2^{-m}[1 + (2^m - 1)(1 - 2p)^{2^m - 1}] - (1 - P)^{2^m - 1} \tag{10.17}$$

where p is the detection probability. Reference [18] also gives a bound on aliasing probability under an independent model. Aliasing probability for compression using LFSR with a primitive feedback polynomial is

$$P_{aliasing} \leq \frac{2}{N + 2} \text{ if } N < N_c$$

$$\leq 1 \text{ if } N = N_c, 2N_c, \ldots \tag{10.18}$$

$$\leq \frac{2}{N_c + 1} \text{ all other values of } N$$

where N_c is the cycle length of the LFSR; for m-degree primitive polynomial, it is equal to $2^m - 1$.

Example 3. The test response from a single output circuit needs to be compressed into a three-bit signature. The LFSR shown in Figure 10.19 is used with feedback polynomial:

$$\phi(x) = x^3 + x + 1$$

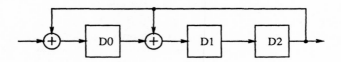

Figure 10.19 LFSR for Example 3.

Table 10.1(a) illustrates the states of the LFSR in Figure 10.19 in response to the input sequence 0101110. One can see that the signature of a fault-free circuit is 010.

Let the good circuit output be represented as a polynomial in x. It can be written as

$$R(x) = 0 \cdot x^6 + 1 \cdot x^5 + 0 \cdot x^4 + 1 \cdot x^3 + 1 \cdot x^2 + 1 \cdot x + 0$$

$$= x^5 + x^3 + x^2 + x$$

Consider the polynomial division $R(x)/\phi(x)$ in modulo-2 arithmetic, such as

$$\frac{x^5 + x^3 + x^2 + x}{x^3 + x + 1}$$

The quotient of the division is $Q(x) = x^2$, and the remainder is $S(x) = x$. In other words, the division can be written as

$$x^5 + x^3 + x^2 + x = (x^2)(x^3 + x + 1) + x$$

Let the test response in the presence of fault f_1 be $R^{f_1} = 0100100$. Table 10.1(b) shows the compression steps. The signature of this response is $S^{f_1} = 010$, which is the same as the fault-free signature. Hence, LFSR compression fails to detect fault f_1.

Let the faulty circuit response in the presence of fault f_2 be $R^{f2} = 0100101$. Table 10.1(c) shows the LFSR contents during the compression of this sequence, where the final signature is $S^{f2}(x) = 110$. Because this is not equal to the good circuit signature, the fault f_2 is detected.

The LFSR compression technique can be extended by using one LFSR per output. However, this will cause large area overhead. Another alternative is to test one output at a time using the same LFSR. The area overhead for this approach is small, but the test time will increase m times. In practice, LFSR compression is extended for multiple output circuits by using an MISR, as shown in Figure 10.20. The feedback structure in the MISR is the same as in a LFSR, except m outputs of the circuit are now connected to the inputs of all m stages of the shift register. MISRs with primitive feedback polynomials are widely used. Some error models for MISR compression and aliasing probability results are discussed in Appendix A.

Table 10.1
The contents of the shift register during syndrome calculation.

Test	Shift	r(x)	Register Values	
t6	1	0	000	
t5	2	1	100	
t4	3	0	010	
t3	4	1	101	
t2	5	1	000	
t1	6	1	100	(a)
t0	7	0	010	
t6	1	0	000	
t5	2	1	100	
t4	3	0	010	
t3	4	0	001	
t2	5	1	010	
t1	6	0	001	(b)
t0	.7	1	010	
t6	1	0	000	
t5	2	1	100	
t4	3	0	010	
t3	4	0	001	
t2	5	1	010	
t1	6	0	001	(c)
t0	7	0	110	

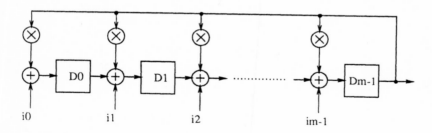

Figure 10.20 MISR compression for a multioutput circuit.

10.9 BIST TEST STRUCTURES

Once the test pattern strategy (pseudorandom, exhaustive, or pseudoexhaustive) and test response compression method (ones count, transition count, and LFSR or MISR) are selected, BIST can be easily implemented. Multiplexers are used to connect the circuit inputs to the test pattern generator, and a test response compressor is connected at the circuit output. Based upon test generation and response compression method, effective fault coverage can be estimated. As it is, the area overhead in adding a test pattern generator and response compressor might be unacceptably high. A number of different circuits, such as registers or scan chains, are used to reduce the hardware overhead.

10.9.1 Built-in Logic Block Observer (BILBO)

In general, large circuits are built in multiple blocks using registers between two stages. The inputs to any combinational block are either the primary inputs or the signals from a register. Similarly, the output of these blocks are either loaded into a register or these are primary outputs. To self-test these blocks, to configure a register at the input of the block as a pattern generator and the register at the output as a response compressor is sufficient. The scan chains are used to feed the appropriate seed vector into the pattern generator and to reset all stages of response compressor. A combinational block can now be tested by a pseudorandom or pseudoexhaustive sequence. After the test application, the signature is shifted out of the output register, again using the scan path. Hence, if the registers in the circuit are designed to be configurable in parallel-load mode (normal circuit operation), scan chain (for feeding the seed and scanning-out the signature), and as ALFSR or MISR (for pattern generation, response compression), the circuit is self-testable. This technique is popularly known as *BILBO* [19].

BILBO uses a register, which can be reconfigured into the following four operational modes (Figure 10.21). The B1, B2 bits are used to select the mode of operation:

1. *B1, B2* = *11*: Configures the register in parallel load mode. This setting is used during normal circuit operation.
2. *B1, B2* = *00*: Converts the register into a linear shift register. This mode is used to feed the seed into a pattern generator, and to scan-out the signature from MISR after the testing is completed.
3. *B1, B2* = *10*: Configures the register into an MISR for test response compression. If the input lines $z_1 z_2 \cdots z_n$ are held at zero, it works as an ALFSR for pattern generation.
4. *B1, B2* = *01*: Reset the register.

BILBO register provides a comprehensive self-test method for large circuits, where the area overhead is low because it can utilize the registers already present on the chip as random pattern generators and signature analyzers.

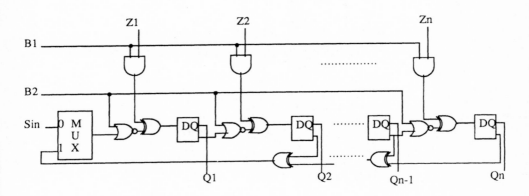

Figure 10.21 An n-bit BILBO register.

10.9.2 Self-Test Using MISRs and Parallel SRSGs (STUMPS)

Figure 10.22, shows a different self-test method used for multichip modules or board testing [20]. This approach uses an extra chip, which contains a *shift register sequence generator* (SRSG), an MISR to compress the output response, and control circuitry to switch between the test and normal modes. All chips in the module are assumed to be designed by using LSSD. The SRSG is used to generate a random pattern. This pattern is shifted in the SRL (scan) chains using the A-shift and B-shift clocks of LSSD. Because the scan chains in various chips may be of different lengths, enough clocks are applied to shift in the longest chain. At the output of the shorter scan chains, some bits of the random sequence are shifted into the MISR. Once all the scan chains are loaded, a system clock is applied and the response of each chip is loaded into its scan chain. This information is compressed in the MISR while the next pattern is loaded. At the end of the test, the signature in the MISR is compared with the fault-free signature.

10.9.3 Circular Self-Test Path

The *circular self-test path* (CSTP) scheme (Figure 10.23) is another design of a self-test method using multifunctional registers [21]. In the self-test mode, a selected subset of registers is converted into a cyclic shift register as shown in Figure 10.23. These registers are modified to act as MISR in the test mode. The input data is XORed with the contents of the previous shift register bit, which is then shifted into the next shift register bit. The circular self-test path is used to generate test patterns, which are applied to all the logic blocks on the chip and the test response is clocked back into the circular self-test path. Because the self-test path works as an MISR, the final content of the MISR is a function of the test vector and circuit response. In the next cycle, the content of the register is used as a test pattern.

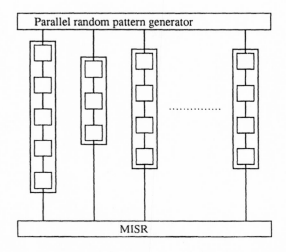

Figure 10.22 Self-test using MISR and parallel SRSG. (From [20], © 1982 IEEE. Reprinted with permission.)

This process of simultaneous test vector generation and compression using the same CSTP is repeated for a number of cycles. Finally, the CSTP contents are examined to determine whether the circuit has a fault.

The main advantage of this method is that all the blocks in the circuit can be tested in one session. The hardware overhead also is low compared to BILBO because the registers must be configured in only two different modes. However, as the same register is simultaneously used for test generation and compression, in certain situations, fault coverage is low. In some circuits, for some seeds, the test patterns repeat after a short sequence, thereby leading to poor testing results.

10.10 SUMMARY

In this chapter, we have discussed design for testability methods. These methods use extra hardware to test the original design. In scan design, all the memory

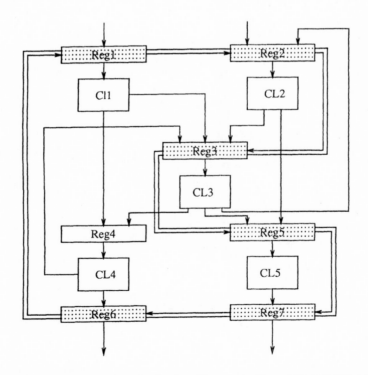

Figure 10.23 Schematic representation of a circular self-test path. (From [21], © 1989 IEEE. Reprinted with permission.)

elements are modified such that they form a shift register during the test mode. A test vector for the combinational block is serially forced into this shift register. The test response is captured in the shift register and serially removed. Variations of this method are available. To reduce the hardware overhead, the partial scan method is used, in which not all flip-flops are included in the scan chain. IEEE Standard 1149.1 boundary scan standards are given for uniformity in scan implementation and to facilitate board testing. To further enhance the observability, a grid-based cross-check technique is given.

Built-in self-test methods are discussed to eliminate the requirement of an external tester. These BIST methods need an on-chip test generator and response compressor. Pseudorandom test generation methods are widely used for BIST. Pseudorandom test generation using LFSR and CA are given. The CAs are easy to build and provide a better test sequence than does LFSR. Different test response compression methods are discussed. Compression methods based on parity, ones count, and transition count are very simple and easy to implement, but they are not accurate. The most widely used compression method is signature analysis with LFSR or MISR. The probability of aliasing in the signature analysis method is discussed. Finally, we have discussed three BIST structures, BILBO, STUMPS, and CSTP.

PROBLEMS

1. Consider the sequential circuit given in Figure 8.10. Modify the D flip-flops in this circuit according to MD-scan. Provide all steps and the test sequence to detect all single stuck-at faults.

2. The architecture of one-bit-slice processor is given in Figure 9.9(a). It has three registers, shifter, register A, and register T. If this slice needs to be designed using partial scan, which register or registers should be selected for the scan chain?

3. The functionality of a sequential circuit is given as follows:

Present State	Next State and Output			
	Input 00	Input 01	Input 10	Input 11
A	C,1	C,0	E,0	E,1
B	D,1	E,0	C,1	C,0
C	B,1	B,0	A,0	A,1
D	E,0	E,1	B,1	B,0
E	A,0	A,1	E,0	E,1

Design this FSM according to LSSD design rules.

4(a). The design of an easily testable PLA is given in Figure 6.5. This PLA has two registers. Implement these registers as LFSRs, and obtain the complete test sequence for this PLA.

4(b). Instead of implementing PLA registers as LFSR, implement them as CA. Compare the testing of this PLA with that of 4(a).

5(a). If the initial condition of an ALFSR is nonzero state, why does it not enter the all-0 state?

5(b). The feedback polynomial of an ALFSR is $[x^4 + x + 1]$. Give the sequence generated by this ALFSR if it initially is in some nonzero state.

6. Consider problem 4(a). Register R in Figure 6.5 is implemented as a BILBO register and register Y as an LFSR. Now, compare the testing with that of problem 4(a) and 4(b).

REFERENCES

1. M.J.Y. Williams and J.B. Angell, "Enhancing Testability of Large Scale Integrated Circuits via Test Points and Additional Logic," *IEEE Trans. Comp.*, **22**(1), pp. 46–60, Jan. 1973.
2. T.W. Williams and K.P. Parker, "Design for Testability—A Survey," *Proc. IEEE*, **71**(1), pp. 98–112, Jan. 1983.
3. E.B. Eichelberger, T.W. Williams, "A Logic Design Structure for LSI Testability," *Proc. Design Auto. Conf.*, pp. 462–468, 1977.
4. R. Gupta, R. Gupta, and M.A. Breuer, "The BALLAST Methodology for Structured Partial Scan Design," *IEEE Trans. Comp.*, **39**(4), pp. 538–544, April 1990.
5. K.T. Cheng and V.D. Agarwal, "A Partial Scan Method for Sequential Circuits with Feedback," *IEEE Trans. Comp.*, **39**(4), pp. 544–548, April 1990.
6. N. Jarwala and C.W. Yau, "A New Framework for Analyzing Test Generation and Diagnosis Algorithms for Wiring Interconnects," *Proc. Int. Test Conf.*, pp. 63–70, 1989.
7. *IEEE Standard 1149.1*, "IEEE Standard Test Access Port and Boundary Scan Architecture," IEEE Press, New York, 1990.
8. T. Gheewala, "CrossCheck: A Cell Based VLSI Testability Solution," *Proc. Design Auto. Conf.*, pp. 706–709, 1989.
9. J. Savir, G.S. Ditlow and P.H. Bardell, "Random Pattern Testability," *IEEE Trans. Comp.*, **33**(1), pp. 79–90, Jan. 1984.
10. P.H. Bardell, W.H. McAnney, and J. Savir, *Built-in Test for VLSI: Pseudorandom Techniques*, John Wiley and Sons, New York, 1987.
11. P.D. Hortensius, R.D. McLeod, W. Pries, D.M. Miller, and H.C. Card, "Cellular Automata Based Pseudorandom Number Generators for Built-in Self-Test," *IEEE Trans. CAD*, **8**(8), pp. 842–859, Aug. 1989.
12. J.A. Waicikauski, E. Lindbloom, E.B. Eichelberger, and O.P. Forlenza, "A Method for Generating Weighted Random Test Patterns," *IBM J. Res. and Dev.*, **33**(2), pp. 149–161, March 1989.
13. E.J. McCluskey, "Verification Testing—A Pseudoexhaustive Test Technique," *IEEE Trans. Comp.*, **33**(6), pp. 541–546, June 1984.
14. P.H. Bardell, "Analysis of Cellular Automata Used as Pseudorandom Pattern Generators," *Proc. Int. Test Conf.*, pp. 762–767, 1990.

15.   S.B. Ackers, "A Parity Bit Signature for Exhaustive Testing," *IEEE Trans. CAD*, **7**(3), pp. 333–338, March 1988.

16.   T.W. William, W. Daehn, M. Gruetzner, and C.W. Starke, "Aliasing Errors in Signature Analysis Registers," *IEEE Design and Test of Computers*, pp. 39–45, April 1987.

17.   M. Damiani, P. Olivo, M. Favalli, and B. Ricco, "An Analytical Model for the Aliasing Probability in Signature Analysis Testing," *IEEE Trans. CAD*, **8**(11), pp. 1133–1144, Nov. 1989.

18.   D.K. Pradhan and S.K. Gupta, "A New Framework for Designing and Analyzing BIST Techniques and Zero Aliasing Compression," *IEEE Trans. Comp.*, **40**(6), pp. 743–763, June 1991.

19.   B. Konemann, J. Mucha, and G. Zwiehoff, "Built-in Logic Block Observation Techniques," *Proc. Int. Test Conf.*, pp. 37–41, 1979.

20.   P.H. Bardell and W.H. McAnney, "Self-Testing of Multichip Logic Modules," *Proc. Int. Test Conf.*, pp. 200–204, 1982.

21.   A. Krasniweski and S. Pilarski, "Circular Self-Test Path: A Low Cost BIST Technique for VLSI Circuits," *IEEE Trans. CAD*, **8**(1), pp. 46–55, Jan. 1989.

Chapter 11
Current Testing

11.1 INTRODUCTION

Test methods to cover stuck-at faults are widely used in the circuit. These methods do not cover physical defects, and thus provide poor coverage of bridging and open faults. In CMOS circuits, the fault detection problem becomes more complex because bridging faults create indeterminate logic outputs. As explained in Chapter 3, even switch-level testing cannot detect CMOS bridging faults. Similarly, as described in Chapter 4, transistor s-open faults create a high impedance state in CMOS circuits. Apart from the difficulty of sensitizing, these faults present severe difficulties in fault-effect propagation. Designing a circuit under scan rules or BIST methods does not solve the problem.

Recently, to overcome this problem, testing methods based on power supply current monitoring have been developed. These methods have been found capable of detecting bridging faults, open faults, some parametric faults such as gate oxide leakage in CMOS circuits. This technique is popularly known as IDDQ testing [1–2]. While conceptually, this technique is very promising, severe instrumentation difficulties have been encountered in its implementations. In this chapter, we will describe the basic concept behind IDDQ testing, its advantages, limitations and available implementation methods.

11.2 BASIC CONCEPT

A CMOS gate consists of an n-MOS pull-down network and a complementary p-MOS pull-up network. For illustration purposes, consider the CMOS NAND gate shown in Figure 11.1(a). In the fault-free steady state, only one part conducts and provides a low impedance path from either GND to the output or V_{dd} to the output. Hence, during steady state, there is no current flow in the circuit except

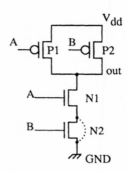

Figure 11.1(a) CMOS NAND gate with a bridging fault to illustrate IDDQ testing.

some junction leakage. The magnitude of this leakage current is on the order of nA and, for all practical purposes, it can be neglected.

Now, consider the presence of a bridging fault. For simplicity, assume the bridging between the source and the drain of n-MOS N2, as shown in Figure 11.1(a). When the gate input is $AB = 10$, the p-MOS P2 is in saturation region. Hence, under this fault, for input $AB = 10$, there is a low impedance path from V_{dd} to GND. The output voltage in this situation is indeterminate, or, in other words, this fault cannot be detected by logic testing. However, due to the presence of low impedance path from V_{dd} to GND, there is a large steady state current in the circuit. The magnitude of this current may be of the order of mA. For this example, the fault-free and faulty currents are shown in Figure 11.1(b). If power supply current is monitored, this large steady-state current is detected, and thus the presence of a fault is known.

Conceptually, this method seems extremely simple. The one point which does not show up in the above discussion is that, during switching from 0-to-1 or 1-to-0, both the *n*-part and the *p*-part conduct. Therefore, during switching, there is a low impedance path from V_{dd} to GND, and thus a large current in the circuit. If the 0-to-1 or 1-to-0 transitions are slow, this switching current will exist for a considerable time. For IDDQ testing, the measurement should be done when this transient current settles down. If a measurement is done before transient current is settled down, a fault-free circuit might be considered faulty under IDDQ testing. Note that the above discussion is limited to one gate. In a VLSI circuit, many gates switch in succession. To an observer, the power supply current, the combined switching current appears as a steady state current in the circuit. Therefore, it is possible that a fault-free circuit is marked as faulty. This phenomenon is illustrated in Figure 11.2, which shows the current in a 4-to-1 multiplexer. The input to this multiplexer is chosen such that four different gates switch in succession. This causes a large current in the circuit for 20ns. If the current measurement is made before

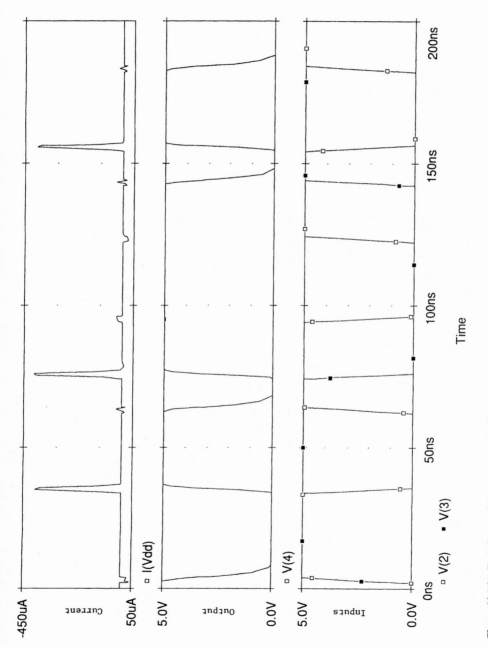

Figure 11.1(b) Fault-free and faulty current for Figure 11.1(a).

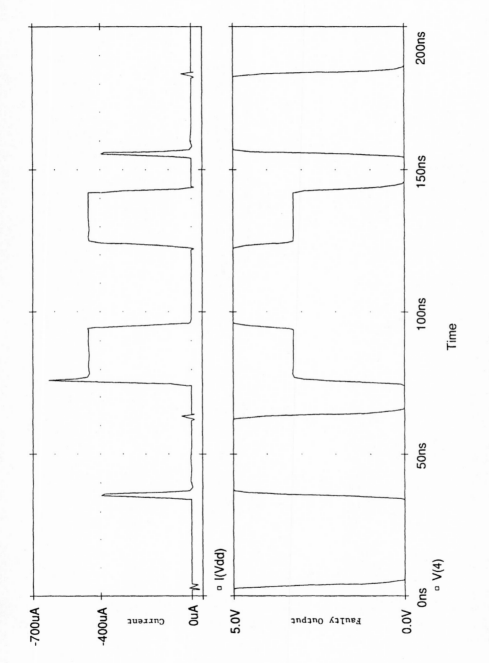

Figure 11.1(b) Fault-free and faulty current for Figure 11.1(a) (continued).

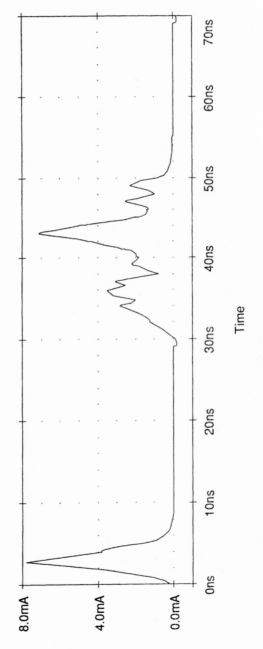

Figure 11.2 Fault-free current in a 4-to-1 multiplexer when four gates switch in succession.

this transient settles, the circuit will be marked as faulty. The circuit used in this example is very small. For large circuits, transients due to successive switching may last up to a few microseconds.

A good estimate of current dissipation in a fault-free circuit is essential for IDDQ testing. If p gates are expected to switch in succession under an input vector, each having an average switching time T_{avg}, the pulse duration for the input signal should be $\gg (p \cdot T_{avg})$ for IDDQ testing to be effective. If this condition is not satisfied, an observer cannot distinguish between fault-free and faulty IDDQ. Also, after estimating the magnitude of the fault-free current, a maximum IDDQ ($IDDQ_{max}$) in the circuit should be specified. A one to two order higher value than $IDDQ_{max}$ can be considered a resolution limit for fault detection.

11.3 ESTIMATION OF FAULT-FREE CURRENT

In recent years, for various reasons, researchers have tried to estimate the fault-free current in the circuit. A good estimate of current in CMOS circuits can be obtained with two factors:

1. *Current dissipation by individual gate during switching.* This includes the magnitude of switching current as well as the duration of switching.
2. *Number of gates that switch their state for a given input vector.* The exact number of gates for each vector cannot be obtained, and hence a statistical distribution is required.

Note that we have completely neglected the leakage current. In a large circuit, this component should also be considered for an accurate estimate.

11.3.1 Current Through a Single Gate

Current through a single gate during switching can be estimated by the dc characteristics of the gate. The rise time as well as the fall time of each gate is estimated with precision during the design. Hence, the time for transient current is known through the design data. This process can be illustrated with the dc characteristics of a CMOS inverter, as shown in Figure 11.3. Originally, the input voltage to the inverter is zero. In this situation, the p-MOS is in saturation and the n-MOS is in the cut-off region. The load capacitance C_L is charged to V_{dd}, there is no conduction path from the V_{dd} to GND, and the current through the circuit is zero.

When the input voltage is more than the threshold voltage V_{tn}, the n-MOS starts conducting, and current through the circuit is no longer zero. Consider the fall time as the time when the output voltage (V_o) drops from ($V_{dd} - V_{tn}$) to $0.1V_{dd}$ (in some VLSI textbooks [3], this time is defined from $0.9V_{dd}$ to $0.1V_{dd}$). At the output node, Kirchhoff's current law dictates:

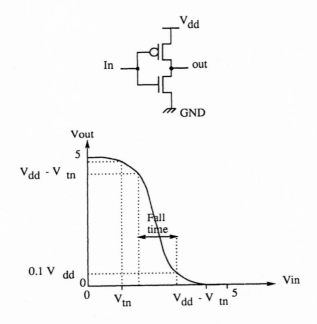

Figure 11.3 A CMOS inverter and its dc characteristics.

$$C_L \frac{dV_o}{dt} + \beta_n \left[(V_{dd} - V_{tn})V_o - \frac{V_o^2}{2} \right] = 0 \tag{11.1}$$

Here, C_L is the load capacitance and β_n is the transistor gain factor, given as

$$\beta_n = \frac{\mu_n \epsilon}{t_{ox}} \left(\frac{W_n}{L_n} \right)$$

where ϵ is the permittivity, μ_n is electron mobility, t_{ox} is gate oxide thickness, W and L are the channel width and length of MOSFET. To calculate the fall time, Equation (11.1) can be integrated from $(V_{dd} - V_{tn})$ to $0.1V_{dd}$:

$$T_{fall} = \frac{C_L}{\beta_n(V_{dd} - V_{tn})} \int_{0.1V_{dd}}^{(V_{dd} - V_{tn})} \frac{dV_o}{\frac{V_o^2}{2(V_{dd} - V_{tn})} - V_o}$$

or

$$T_{fall} = \frac{C_L}{\beta_n(V_{dd} - V_{tn})} \ln\left[\frac{19V_{dd} - 20V_{tn}}{V_{dd}}\right] \tag{11.2}$$

If we approximate $V_{tn} \approx 0.2V_{dd}$, for a 5 V process, Equation (11.2) reduces to

$$T_{fall} = \frac{0.66\ C_L}{\beta_n} \tag{11.3}$$

Similarly, the approximate rise time is given as

$$T_{rise} = \frac{0.66\ C_L}{\beta_p} \tag{11.4}$$

Equations (11.3) and (11.4) give the approximate time duration of current in the circuit when the circuit switches from 1-to-0 and 0-to-1, respectively. The magnitude of the current during these periods can be approximated by the linear region current of the n-MOS and the p-MOS, respectively. The peak magnitude of the current is the saturation current through the n-MOS and the p-MOS. The current through n-MOS is given as

$$I_{peak} = \frac{\beta}{2}(V_{in} - V_{tn})^2 \tag{11.5}$$

$$I_{avg} = \beta\left[(V_{in} - V_{tn})V_o - \frac{V_o^2}{2}\right] \tag{11.6}$$

Using an equivalent inverter model, the magnitude of switching current through any gate as well as switching time can be estimated. For an n-input NAND gate, the rise and the fall times can be approximated as:

$$T_{rise} = \frac{0.66\ (nC_D + C_L)}{\beta_p} \tag{11.7a}$$

$$T_{fall} = \frac{0.66n\ (nC_D + C_L)}{\beta_n} \tag{11.7b}$$

Here, C_D is the capacitance of drain area of a unit size n-MOS transistor. The rise and the fall times for a NOR gate can be approximated as

$$T_{rise} = \frac{0.66n \ (nC_D + C_L)}{\beta_p} \tag{11.8a}$$

$$T_{fall} = \frac{0.66 \ (nC_D + C_L)}{\beta_n} \tag{11.8b}$$

Note that these estimates of the rise and fall times alone do not serve our purpose. To know how many gates will change their state each time the values at the primary inputs are changed is essential.

11.3.2 Estimation of Current in a Circuit

The current in the circuit can be estimated by the number of gates that will switch their state if the input to the circuit is changed. Unfortunately, the number of gates that will switch their state is not fixed and, in general, one cannot estimate this number deterministically. In CREST [4–6], the switching of a gate is considered as a stochastic process to obtain a probability density function. Based upon this probability, using event-driven simulation, the expected current and its variance are estimated. The basic idea is to consider the events at the input of a gate to derive an event at its output. Every time an event takes place at the output of a gate, an expected current pulse (equivalent to the gate switching current) is added to the global expected current.

If j_k is the height of total current in the circuit, where $k = 1, 2, \ldots, m$ represents a train of pulses, and t_0 is the time duration of a pulse, the effective current in the circuit can be given as [4]:

$$J_{eff} = \sum_{k=1}^{N} J_k \cdot P_k \tag{11.9}$$

where J_k is defined as:

$$J_k = \frac{1}{t_0} \int_0^{t_0} f(j_k) \cdot dt$$

where f is a nonlinear function, based upon [6], as shown in Figure 11.4. Note that

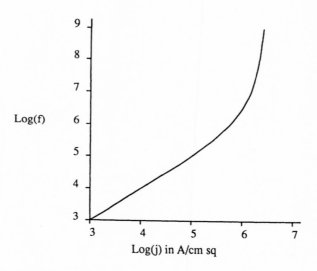

Figure 11.4 Plot of function f, based upon [6].

the probabilities P_k can be calculated as described in Chapter 5, Section 5.4. Equation (11.9) can be rewritten as

$$J_{eff} = \sum_{k=1}^{N} \left[\frac{1}{t_0} \int_0^{t_0} f(j_k) \cdot dt \right] \cdot P_k = \frac{1}{t_0} \int_0^{t_0} \left[\sum_{k=1}^{N} f(j_k) \cdot P_k \right] \cdot dt \tag{11.10}$$

or

$$J_{eff} = \frac{1}{t_0} \int_0^{t_0} E[f(j)] \cdot dt \tag{11.11}$$

where E denotes the expected value. Using this expected value, the variance of the current can be given as:

$$\text{Var}[j(t)] = E^2[j(t)] - \{E[j(t)]\}^2 \tag{11.12}$$

In CREST, the expected current pulse per transition is assumed to be a triangular pulse that starts with a maximum value at the time of transition and linearly decays to zero. If C_p and C_n are the lumped capacitances between the output and V_{dd} and between the output and GND, respectively, the expected current in CREST is approximated as

$$E[j(t)] = E[j_p(t)] \frac{C_n}{C_n + C_p} + E[j_n(t)] \frac{C_p}{C_n + C_p} \qquad (11.13)$$

As it is explained in the previous section, the time duration (t_0) for this pulse is the rise-time or the fall-time of a gate. This time can be considered as an average switching time, i.e., ($T_{rise} + T_{fall}$)/2. In CREST, the magnitude of current is approximated by that flowing through the lumped capacitances at the output, as given in Equation (11.13). However, as explained in the previous section, the maximum magnitude of the current pulse is the saturation current through a MOSFET in the equivalent inverter model, while the average value can be approximated as a linear region current.

The motivation for developing CREST was to estimate the electromigration failure rate. However, this technique is equally useful to estimate fault-free current for IDDQ testing. After an estimate of fault-free current, using an external or on-chip current sensor, the power supply current can be monitored and a high current or faulty state can be identified.

11.4 CURRENT SENSING TECHNIQUES

During a normal logic level testing, power supply current can be monitored, and if this current is found higher than the specified expected current, the circuit can be identified as faulty. In concept, this testing method is simple. However, as mentioned earlier, the measurement of power supply current is not straightforward. A typical VLSI circuit I/O pins may have about 100 Ω output impedance, driving a 50 pF, 100 Ω line [7–8]. Simultaneous switching of multiple pins may draw up to 5 A current transient of about 10 ns pulsewidth with an edge speed of 10 A/ns. A simple current probe offers significant loading at the power supply, causes a large voltage drop across it, and is lacking in dc accuracy. A possibility is to use an elaborate measurement method to bypass the transients and to compensate for the voltage drop. A second possibility is to build an on-chip current sensor as part of the BIST. In this section, we describe both external and on-chip current sensing methods.

11.4.1 External Current Sensor

Some ac as well as dc current probes are commercially available. In [8], a comprehensive discussion is given on the use of external current probes. Conceptually, a current probe can be used between the circuit-under-test and the power supply, as shown in Figure 11.5. The basic problem with such probing is the insertion inductance, which is typically 10 to 50 nH. A current pulse of edge speed about 10 A/ns fed to a 10 nH inductive probe causes about a 100 V voltage drop across it. Thus, such probes cannot be used for IDDQ testing.

A simple solution is to use an *operational amplifier* (op amp) with sufficient gain, while keeping the current sense resistor in its feedback loop. Schematically, it is shown in Figure 11.6(a). This op amp should be designed to compensate the voltage drop across the sense resistor as well as be able to supply high transient currents to the circuit. Obviously, designing such a circuit is very difficult and costly.

The solution of this problem is to provide a shunt path for the transient current across the sense resistor. If this shunt path is provided by a diode (Figure 11.6(b)), it still would cause about a 0.6 V drop across it, and hence cannot be used in production testing due to the specified testing voltage. To avoid this voltage drop across the shunt path, a FET bypass circuit is used (Figure 11.6(c)). This bypass transistor is on only during transient. Thus, when the transient settles down, the current passes through the sense resistor. To filter the high impedance noise at high frequencies, a small capacitor is added between the sense circuit and the circuit-under-test, as shown in Figure 11.6(d). It has been reported that a 2000 to 2500 pF capacitor and a 400 to 500 Ω resistor provide an adequate bypass circuit.

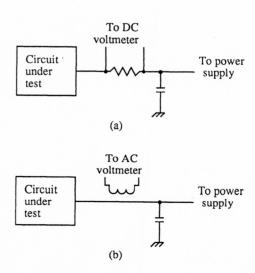

(a)

(b)

Figure 11.5 Schematics of current measurement technique: (a) ac current probe; (b) dc current probe. (From [8], © 1987 IEEE. Reprinted with permission.)

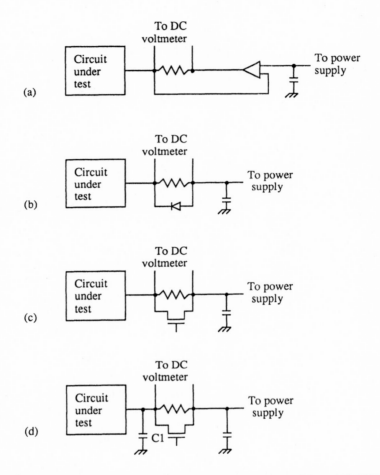

Figure 11.6 Current measurement techniques using external probe. (From [8], © 1987 IEEE. Reprinted with permission.)

The only disadvantage with this method is that it causes an RC loading at the output. Hence, the circuit takes more time to stabilize. However, it can still be tested at about 100 kHz test frequency.

If the circuit in Figure 11.6(d) is examined critically, one may realize that the resistance in the sense circuit is redundant. The only requirement in current sensing is a bypass circuit for the transient currents. By eliminating this resistance, the testing speed can be significantly improved. This modification is shown in Figure 11.7. In this circuit, as before, the FET is on during the transient state when the

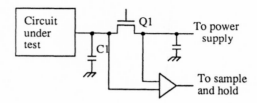

Figure 11.7 External current sense circuit for fast response time. (From [8], © 1987 IEEE. Reprinted with permission.)

circuit-under-test is drawing a large current. Once transients settle down, the FET is off and capacitor C_1 supplies the static current to the circuit-under-test. The I_{dd} is measured by the voltage drop across the FET. Note that in this circuit the value of capacitor C_1 is critical. It should be chosen such that it will keep the voltage (V_{dd}) to the specified testing voltage in the fault-free circuit, at least until the measurement is done.

Example 1. Consider the transient of 1 A peak current which lasts 5 ns. A desired resolution for I_{dd} is 100 μA per 10 mV. The desired I_{dd} measurement time is 500 ns. For these specifications, the value of C_1 can be estimated as C_1 = (100 μA · 500 ns)/10 mV = 5000 pF.

Suppose the FET is turned off at time t = 0 ns after the transients settle down. The voltage across it is measured as 1 V after 100 ns. The charge delivered by C_1 is Q = (1 V) · (5000 pF) = 5 nC. This 5 nC charge is delivered in 100 ns, which means that the circuit drew a static current of 50 mA during 100 ns. Although the value may change, depending upon circuit, in general, this current is higher than the expected current. Hence, the circuit has a fault.

The application of test vectors at a 10 MHz rate, as considered in the above example, is a little optimistic. A 100 kHz to 1 MHz testing frequency is more realistic for IDDQ testing with an external current sensor. If the logic testing is done at this rate, IDDQ testing can be done simultaneously. However, it will cause an increase in test time by an order of one to two.

11.4.2 Built-in Current Sensor

The external current probe in IDDQ testing is costly equipment and its calibration to the desired resolution is difficult. Furthermore, external current sensing can only be done at slow speed, and thus the test time increases significantly. The testing time can be improved by eliminating the external current probe and using an on-chip current sensor. This possibility was examined in [9]. In this approach, the whole circuit is divided into several sections, and each section is associated with a current sensor. This concept is shown in Figure 11.8(a). A current sensing circuit, made up of a circuit breaker and differential amplifier, is built on-chip with every section to detect abnormally high currents. The circuit schematics of this current sensor is shown in Figure 11.8(b). In the fault-free situation, the virtual ground is logic 0 and the potential at node 2 is about 5 V. Thus, under the fault-free situation, transistor T_1 is on and T_2 is off.

In the presence of a fault, when it is sensitized, the circuit-under-test draws high current. As the voltage at virtual ground increases, transistor T_2 switches on and T_1 switches off. This effectively isolates the circuit-under-test from the power supply. Note that a third transistor T_3 is used to ensure that the circuit breaker will operate in the correctly conducting state. The purpose of this transistor is to restore the voltage at node 3. Hence, it is designed to offer a high on resistance and allows only a small leakage current under the fault-free situation.

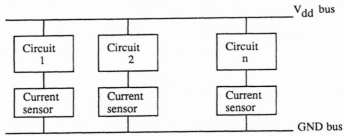

Figure 11.8(a) Schematic representation of the use of on-chip current sensor for IDDQ testing.

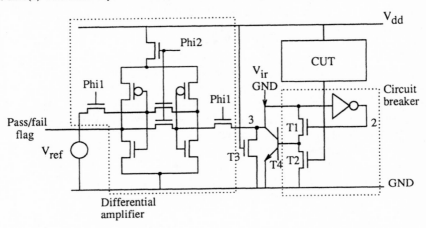

Figure 11.8(b) Circuit schematics of on-chip current sensor. (From [9], © 1988 IEEE. Reprinted with permission.)

The differential amplifier compares the virtual ground voltage with a reference voltage. This reference voltage is predetermined on the basis of virtual ground voltage induced by the normal static current through the circuit-under-test. This differential amplifier is carefully designed to achieve the required switching resolution and to minimize the amplifier's offset sensitivity. The output of this amplifier is a pass–fail flag, identifying a fault-free–faulty circuit.

The major advantage of this approach is that IDDQ testing can be implemented as a BIST. The on-chip current sensing avoids the large inductive circuit. Thus, the on-chip IDDQ testing can be performed at much higher speeds than in external current sensing.

The major problem with an on-chip current sensor is that it requires the whole circuit to be partitioned into several sections. Design of the current sensor and partitioning the circuit significantly complicate the chip design. This problem limits the use of on-chip current sensors. Furthermore, at the present time, the size of on-chip current sensors is large and causes significant area overhead. This sensor also permanently loads the power supply buses, and hence causes a penalty in normal circuit operating speed. External current sensing does not have these problems, and at present IDDQ testing is mostly based on external current sensing.

11.5   TEST GENERATION FOR IDDQ TESTING

To generate test vectors for IDDQ testing, a switch-level representation is necessary. In Chapter 5, Section 5.5, a switch-level test generation method is given for combinational circuits. In this method, every logic gate is represented as a connection graph. All possible paths from the V_{dd} to the output in the p-graph, and from the output to GND in the n-graph are identified and examined. If there are two or more paths in parallel, only one path is sensitized at a time. All other paths are kept off by applying a 0 to at least one edge. To test for a shorted edge in the n-graph (or p-graph), a 0 (1) is applied to that edge while 1s (0s) are applied to all other edges in that path. Note that, in the presence of a fault, there is a path from the V_{dd} to GND and the circuit draws a large static current. This current is monitored by an on-chip or external current sensor. Hence, for IDDQ testing, this switch-level test generation method is sufficient.

Although not mentioned explicitly in Section 5.5, this test generation method is quite suitable for VLSI circuits with complex gates. The standard gate level test generation methods are not applicable for circuits with complex gates. Standard gate-level test methods use an equivalent AND-OR model to represent a complex gate, and hence these methods cannot provide a test vector for an internal bridging fault, and thus are not suitable for IDDQ testing.

Note that the switch-level test generation method discussed in Section 5.5 provides all possible test vectors. IDDQ testing does not require applying all possible test vectors. The only objective is to examine all possible paths from the V_{dd} to GND. Hence, to make this test generation effective for IDDQ testing, don't-cares are eliminated by assigning specific 1s and 0s, and then redundant vectors are deleted from the test set. The rules to replace a don't-care (x) by a 1 or 0 are as follows:

Rule 1: If an input to a NOR gate is x, replace it with a 0.

Rule 2: If an input to a NAND gate is x, replace it with a 1.

The philosophy behind the above two rules is to eliminate unknown values so that deterministic fault coverage can be obtained. A NOR (NAND) gate has multiple edges in series in the p-graph (or n-graph). Thus, a shorted edge can be detected if the edge under consideration has a 1 (0), while other edges have 0 (1). In the presence of a shorted edge or transistor stuck-on fault, the p-graph (or n-graph) appears as shorted, and hence the fault is detected.

Example 2. Using this method, if we examine the switch level test set for Figure 5.18 (reproduced for convenience as Figure 11.9), the complete test set is $ABCEF$ = [111xx, 11x00, 1101x, 110x1, 0x1xx, x01xx, 0xx00, x0x00]. In this circuit, gates G_1, G_3, and G_4 are the NOR gates and G_2 is a NAND gate. Thus, any x at A or B is replaced by a 1, and an x at C, E, F, D, L, and M is replaced by a 0. Thus,

$$ABCEF = [11111, 11100, 11011, 11011, 00111, 00111, 00100, 00100]$$

After deleting the multiple entries, the final test set is given as

$$ABCEF = [11111, 11100, 11011, 00111, 00100]$$

Note that this circuit has five primary inputs. An exhaustive test set for this circuit is 32 test vectors. The test set obtained above is only five vectors long. These five vectors are sufficient to detect all bridging and transistor stuck-on faults under the IDDQ test environment. As discussed in section 5.5, these five vectors also detect all single stuck-at faults under logic test environment.

The above example does not include a complex gate. For test generation, a complex gate is modeled as a *basic cell matrix* (BCM) [10]. The BCM models for sum-of-product and product-of-sum forms are given in Figure 11.10. The test vectors for the complex gate using these models are obtained as given in Figure 11.11. Note that these sequences are given only for the n-part under the single-fault assumption, the test sequences for the p-part can be obtained by complementing the values.

1. In the BCM model of Figure 11.10(a), a transistor stuck-on fault or bridging fault between the source and drain of a transistor, and a bridging fault between two internal nodes can be detected by the sequence T_1. In the same circuit,

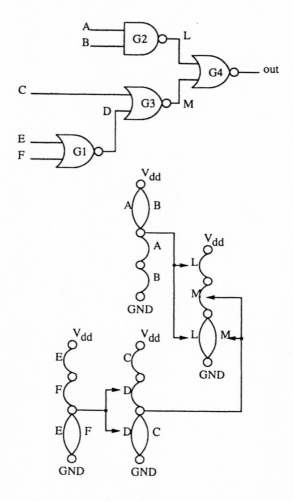

Figure 11.9 A sample combinational circuit with its graph to generate test vectors at the switch level.

a bridging fault between two logical inputs can be detected by the sequence T_2.

2. In the BCM model of Figure 11.10(b), a bridging fault between two logical inputs can be detected by the sequence $[T_2 + T_3]$. In the same circuit, a transistor stuck-on fault or bridging fault between the source and drain of a transistor and a bridging fault between any two internal nodes can be detected by the sequence T_4.

Example 3. Consider the CMOS complex gate shown in Figure 11.12. This complex gate has four primary inputs. Hence, an exhaustive test set for this gate is 16 test

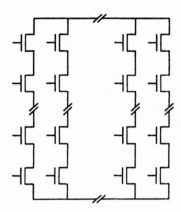

Figure 11.10(a) BCM model (type A) for sum-of-product gate.

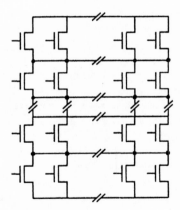

Figure 11.10(b) BCM model (type B) for product-of-sum gate.

vectors. The *n*-part of this gate can be modeled as BCM type B, and the *p*-part can be modeled as BCM type A. Using the above method, the test set to detect all bridging and transistor stuck-on faults under IDDQ testing can be given as

$$ABCD = [1001, 0110, 0011, 1100]$$

Some results for IDDQ testing are summarized as follows:

1. All single nonredundant bridging faults in a combinational circuit are detected

by IDDQ testing [11]. A test vector that detects a single bridging fault f_i under IDDQ testing also detects all multiple faults which contain f_i.

2. A test set that detects all single p-MOS stuck-on faults also detects all single line s-a-0 faults. A test set that detects all single n-MOS stuck-on faults also detects all single line s-a-1 faults. Hence, testing a static CMOS circuit for all transistor stuck-on faults is sufficient to cover all line stuck-at faults.

3. The test procedure as described above provides 100% bridging and transistor stuck-on fault coverage under IDDQ testing. It also provides 100% single line stuck-at fault coverage under logic testing.

11.6 SUMMARY

IDDQ testing, in general, is a very promising method. The advantages are enormous. A very small test set provides extremely good coverage of not only physical defects such as bridging faults, but also potential reliability problems such as gate oxide leakage. Measurement of current at an elevated voltage and temperature (stress testing) is already considered an essential test and is used in industry for reliability screening. The exact details of such testing are given in Chapter 12.

The basic advantage of IDDQ testing is that this technique does not require propagation of the fault effect to the primary outputs. As soon as a fault is sensed, the fault effect is observed at the power supply pin. As an input pattern sensitizes a path from the primary inputs to the primary outputs, each vector covers a large number of faults. Apart from single-fault coverage, IDDQ testing also covers all multiple faults. A multiple fault is not masked in IDDQ testing. Standard logic testing, in general, cannot detect bridging faults in CMOS circuits. IDDQ testing can detect any bridging faults between any two internal nodes of a gate. This includes bridging between two nodes, one in the *n*-part and the other in the *p*-part. In the literature, this technique has also been reported to be effective in detecting stuck-open faults.

While this technique has enormous advantages, precautions are necessary in its use. Some circuits inherently have high current states and in some other circuits these states occur unintentionally. Such high current states must be identified and masked during IDDQ testing. During IDDQ testing, a precaution must be taken such that a vector does not produce a high impedance gate at the I/O pins. Finally, the normal current dissipation should be estimated with care and resolution for fault detection should be kept at least one-to-two orders of magnitude higher.

The major disadvantage of IDDQ testing is the relatively slow measurement speed. Measurements done at MHz rates are effective when the output power buffers have a tristate design. In external current sensing, the tester impedance creates a transient in the microsecond range. In this situation, fast measurement

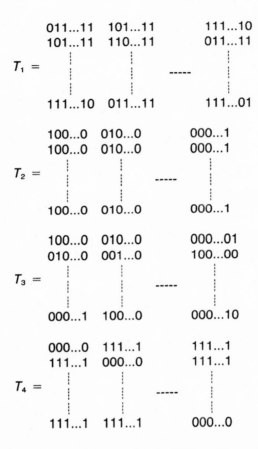

Figure 11.11 Test sequences for a complex gate using the BCM model.

would lead to the wrong conclusion. The on-chip current sensors cause hardware overhead, a penalty in normal circuit speed, and, most importantly, complicate the chip design due to the partitioning requirement. Due to these drawbacks, at the present time, on-chip current sensors have been used in laboratories, but not in production testing.

For effective testing, the test generation must be done at the switch level. The standard gate-level test methods are not suitable for internal bridging faults. At the present time, the switch level test methods are not used in production testing. Hence, this can also be considered as a limitation of IDDQ testing.

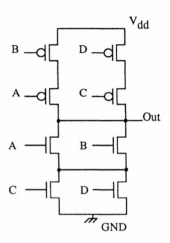

Figure 11.12 CMOS complex gate for Example 3.

PROBLEMS

1. Consider a three-input CMOS NAND gate with the following parameters: $\beta_n = 40 \ \mu A/V^2$ and $\beta_p = 20 \ \mu A/V^2$ ($W = 12 \ \mu m$ and $L = 4 \ \mu m$). For five-volt power supply, calculate the approximate rise time and fall time. If a complex gate is made with the same transistor parameters to implement $\overline{AB(C + D)}$, what will be the approximate rise time and fall time?

2. Estimate the value of peak current for the NAND and complex gate of problem (1) during 1-to-0 and 0-to-1 transitions.

3. A circuit is put to IDDQ test using external current sensor as given in Figure 11.6(d), with $C_1 = 5000$ pF. During switching, circuit transients are expected to last for 10 ns with peak current 3 A. During steady state, the circuit is supposed to draw a 10 μA static current. The voltage across bypass FET is sampled at every 200 ns. If the voltage across the bypass FET is measured as 0.8 V for a vector, determine whether or not the circuit has a fault.

4. Generate a minimal test set at the switch level for IDDQ testing of the circuit given in Figure 5.12(a).

REFERENCES

1. Y.K. Malaiya and S.Y.H. Su, "A New Fault Model and Testing Technique for CMOS Devices," *Proc. Int. Test Conf.*, 1982, pp. 25–34.

2. C.F. Hawkins, J.M. Soden, R.R. Fritzemeier, and L.K. Horning, "Quiescent Power Supply Current Measurement for CMOS IC Defect Detection," *IEEE Trans. Industrial Electronics*, **36**(2), May 1989, pp. 211–218.

3. N. Weste and K. Eshraghian, *Principles of CMOS VLSI Design*, Addison Wesley, Reading, MA, 1985.

4. R. Burch, F. Najm, P. Yang, and D. Hocevar, "Pattern Independent Current Estimation for Reliability Analysis of CMOS Circuits," *Proc. Design Auto. Conf.*, pp. 294–299, 1988.

5. F. Najm, I. Haji, and P. Yang, "Computation of Bus Current Variance for Reliability Estimation of VLSI Circuits," *Proc. Int. Conf. CAD*, pp. 202–205, 1989.

6. F. Najm, I. Haji, and P. Yang, "Electromigration Median Time-to-Failure Based on a Stochastic Current Waveform," *Proc. Int. Conf. Computer Design*, pp. 447–450, 1989.

7. C. Crapushettes, "Testing CMOS IDD on Large Devices," *Proc. Int. Test Conf.*, pp. 310–315, 1987.

8. M. Keating and D. Meyer, "A New Approach to Dynamic IDD Testing," *Proc. Int. Test Conf.*, pp. 316–321, 1987.

9. D.B.I. Feltham, P.J. Nigh, L.R. Carley, and W. Maly, "Current Sensing for Built-in Testing of CMOS Circuits," *Proc. Int. Conf. Computer Design*, pp. 454–457, 1988.

10. R. Rajsuman, A.P. Jayasumana, and Y.K. Malaiya, "Testing of Complex Gates," *Electronics Letts.*, **23**(16), pp. 813–814, July 1987.

11. K.J. Lee and M.A. Breuer, "On Detecting Single and Multiple Bridging Faults in CMOS Circuits Using the Current Supply Monitoring Method," *Proc. Int. Symp. Circuits and Systems*, pp. 5–8, 1990.

Chapter 12
Reliability Testing
By K. Rajkanan* and R. Rajsuman

12.1 INTRODUCTION

A user specifies a minimum average fraction of the purchased ICs which must pass an agreed-upon specification test, an *average quality level* (AQL). From a user's perspective, not only initial functionality with a minimum quality level is important, but also the probability of successful performance of the required functions under stated conditions for a stated period of time, i.e., reliability. Thus, while the functional testing is crucial, the reliability testing is equally important.

Within the agreed-upon specifications and usage, many factors affect the quality and reliability of an IC. The quality and reliability depend upon the physical phenomena occuring due to design, workmanship defects, wear-out, and overstress. Although many factors affecting the quality and reliability can be measured and associated with a physical phenomenon, there is no inherent limitation on improving the quality and reliability. In principle, every physical cause that adversely affects the quality can be eliminated if there is an economic justification. The overall quality and reliability cannot be predicted with precision due to statistical variations, as well as human factors and the perception of quality and reliability. Yet the motivation for quality and reliability testing and prediction is obvious. Definitely, they are needed to make sound economic decisions during design and manufacturing, and forecasting items such as system cost, operational effectiveness, warranty cost, logistics support, and competitive position. In this chapter, we will describe the factors affecting the intrinsic quality and reliability of integrated circuits and methods to measure these parameters.

---

*K. Rajkanan is with Zilog, Inc.

12.2 COMPONENT QUALITY AND FAULT COVERAGE

The quality of an IC is judged by the AQL, which is given as

$$\text{AQL} = \frac{\text{defective ICs received}}{\text{total number of ICs received}} \tag{12.1}$$

AQL depends on several parameters, many of which are dependent upon human factors such as mislabeling and mishandling, leading to bent pins. In addition, the fault coverage, which is the ratio of all detected faults to all possible faults, also affects the AQL.

To study the effect of fault coverage on AQL, a test program with high fault coverage (such as 99.9%) is divided into several test sets with varying fault coverage, for example, 50%, 60%, and 70% fault coverage. A number of circuits which are known to pass full test are then retested with test programs of varying fault coverage. In this way, AQL can be calculated for different fault coverage. The data yield a lower bound on the relationship between fault coverage and quality. In general, the data give a straight line on a log-log plot (Figure 12.1). As expected, high fault coverage is needed for a low AQL. If the manufacturing process is extremely good and only few defects are expected, the test set should do very well in detecting individual faults. For constant defect density, smaller dies are less prone to defects

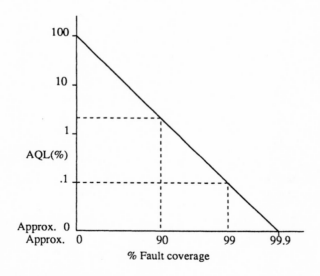

Figure 12.1 Relationship between AQL and percentage fault coverage.

as compared to the larger chips. Therefore, low fault coverage can be tolerated for a given AQL on smaller chips.

12.3 RELIABILITY AND FAILURE RATE

Reliability of a system may be stated in terms of system failure rate or *mean time between failures* (MTBF). For a constant failure rate, MTBF is the inverse of failure rate (λ), and is given by

$$\text{MTBF} = \frac{1}{\lambda} \tag{12.2}$$

Semiconductor chips are used in large quantities. Therefore, failure per unit time is more convenient to describe the reliability. A common measure to describe IC failure rate is the *failure unit* (FIT), defined as

$$1 \text{ FIT} = \text{one failure in } 10^9 \text{ device hours} \tag{12.3}$$

IC reliability is often characterized by a bathtub curve, as shown in Figure 12.2. This curve illustrates a relatively high failure rate during the infant mortality phase, comparatively low and constant failure rate during the life phase, and increasing failure rate during the wear-out phase.

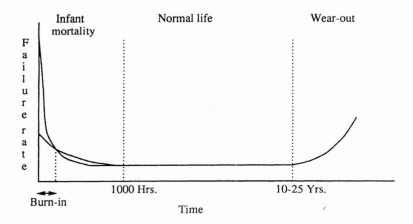

Figure 12.2 Schematic of IC failure rate characteristics.

The infant mortality region depends upon the maturity of the manufacturing and packaging processes. As shown in Figure 12.2, burn-in under high temperature and stress voltage accelerates the infant mortality and flattens the curve. This reduces the field failure, and thus enhances the reliability over the IC's useful life. The constant failure rate region is the useful life of an IC. The failure rate can be estimated by the constant failure rate of components multiplied by the number of active components. The failure in this region is random in nature and may be caused by a mixture of manufacturing and design defects requiring a long period to show up. Some examples of such faults are tolerance mismatch between device parameters, timing problems, and parameter drift.

In the wear-out region, the failure rate continuously increases with time. Systems are designed such that the components do not approach the wear-out region within the stated lifetime. Extreme conditions, such as high radiation, high current or voltage stress, and severe environmental conditions may cause wear-out. System life expectancy can be judged by performing the tests under high stress conditions, which accelerate the onset of component wear-out failures.

In order to predict reliability at a future time or some specific operating conditions, the circuit lifetime distribution is described as a function of time and operating conditions. Generally, four different failure probability density functions, *normal*, *log-normal*, *Weibull* (log-log), and *exponential* distributions are used. These functions are schematically shown in Figure 12.3.

Normal and exponential failure probability density functions are the simplest in terms of mathematical formulation and may adequately describe the wear-out phase and constant failure rate regions. In general, the failure of an IC is a result of several random variables interacting multiplicatively, producing either a log-normal or log-log distribution. Although the Weibull distribution is shown to be effective in describing infant mortality, the log-normal distribution is used for a wide range of failure mechanisms.

For a given median life (time to fail 50% of the tested units), the failure rate (in FIT) depends on the standard deviation of life distribution curve, σ. Determination of the instantaneous failure rate of an IC does not provide an accurate measure. Therefore, the average cumulative failure rate is used to describe the failure rate. Cumulative failure rate is given as

$$\lambda_{cum} = \frac{F(t)}{(n \cdot t)} \tag{12.4}$$

where t is the cumulative test time, $F(t)$ is the cumulative failures at time t, and n is the number of ICs put to life test.

Because of the sampling variance, the use of a single point estimate might result in a significant error with respect to the true failure rate. Also, to repeat the

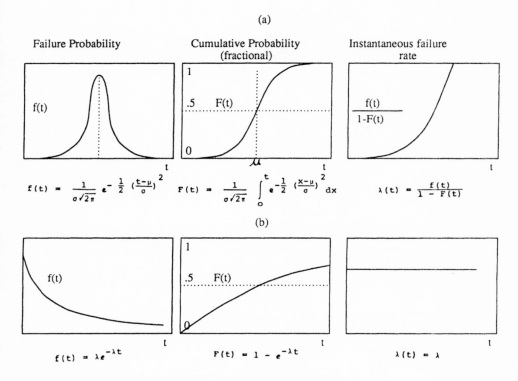

Figure 12.3 Normal (a) and exponential (b) failure probability functions.

reliability experiment many times is not practical to determine the confidence interval of the cumulative failure rate for given test conditions. However, the median life follows the central limit theorem of statistical phenomena. Therefore, the chi-square distribution is used to estimate the maximum and minimum cumulative failure rate with certain confidence level, β [2]. The maximum and minimum cumulative failure rates after time t are given by

$$\text{Maximum Cumulative Failure Rate} = \frac{\chi^2(df, \alpha)}{2nt} \tag{12.5}$$

$$\text{Minimum Cumulative Failure Rate} = \frac{\chi^2(df, \beta)}{2nt} \tag{12.6}$$

where χ^2 is the chi-square value, df is the number of degrees of freedom given as

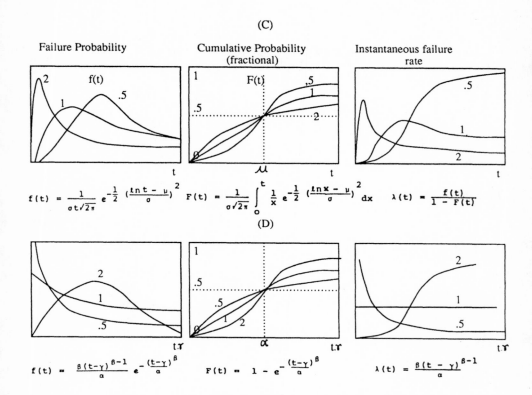

Figure 12.3 (cont.) Log-normal (c) and Weibull (d) failure probability functions.

$(2 \times$ number of failures $+ 2)$, $\alpha = 1 - \beta$, where β is the upper confidence limit, n is the number of units under test, and t is the cumulative test time.

Tables for chi-square values can be found in statistics textbooks. For large df, the χ^2 distribution is related to the normal distribution:

$$\chi^2 = df\left[\left(1 - \frac{2}{9\,df}\right) + Z_p\cdot\left(\frac{2}{9\,df}\right)^{1/2}\right]^3 ; \text{ for } df > 30 \tag{12.7a}$$

$$\chi^2 = 0.5[Z_p + (2\,df - 1)^{1/2}]^2; \text{ for } df > 100 \tag{12.7b}$$

Here Z_p is a Gaussian random variable, the cumulative distribution function of which is β. Values of Z_p for different confidence limits are given as

Confidence Level	Value of Z_p
60%	0.254
70%	0.524
80%	0.842
90%	1.282
95%	1.645

Example 1. Suppose the current IC failure rate is 10 FIT. We want to predict the failure rate with an upper confidence limit of 60% from single failure. Using Equation (12.5), $df = 4$, $\alpha = 1 - \beta$ is 0.04; so $\chi^2 = 4.04$. This implies that we have to study about 200 million device hours to predict the failure rate with 60% upper confidence level. More chip hours are needed for a higher confidence level.

The above example clearly shows the impracticality of this procedure. Accelerated test methods are used for reliability testing and prediction. A detailed discussion of accelerated test methods is given in Section 12.6.

12.4 FAILURE MECHANISMS

Various physical and chemical processes may cause a failure. The cause of failure is termed the *failure mode*, and the physical or chemical process which causes the failure is termed the *failure mechanism*. For example, electromigration is a failure mechanism that leads to the failure mode of an open circuit. Similarly, the ionic movement in the oxide is a failure mechanism that leads to a failure mode of drift in threshold voltage.

Failure mechanism can be broadly divided into four categories: (1) *chip related*, (2) *assembly related*, (3) *operation induced*, and (4) *application induced*. In Table 12.1, various failure mechanisms are listed along with the stress which aggravates that particular failure mechanism.

12.4.1 Chip Related Failures

Chip related failures can be categorized into (1) *dopant related failure*, (2) *oxide related failures*, (3) *metallization related failures*, and (4) *chip design related failures*.

Undesirable diffusion of the dopant may occur due to lattice defect sites caused by the thermal processes. Nonuniform current may flow in the circuit due to this undesirable diffusion. This may lead to low punch-through voltage and an increase in junction leakage.

High temperature helps in releasing the available mobile ions in the transistor. These mobile ions might have been trapped in silicon during manufacturing, or

Table 12.1
Some failure mechanisms and corresponding aggravating stress.

Fault Type	Failure Mechanism	Aggravating Stress
Chip Related	Dopant diffusion	High temperature
	Ionic movement	High temperature, voltage and humidity
	Oxide contamination	High temperature and voltage
	Dielectric breakdown	High temperature and voltage
	Junction spiking	High temperature
	Metal voiding	High temperature and current
	Metal interconnect cracking	High temperature
	Chemical corrosion	High temperature, voltage and humidity
Assembly Related	Mounting voids	High temperature
	Thermal mis-match	Temperature cycling
	Alloy segregation	Power cycling
	Intermetallic growth	High temperature
	Metal stress relaxation	High temperature
	Fatigue stress	Temperature cycling and vibration
	Encapsulation aging	High temperature and humidity cycling
	Condensation	Low temperature
	Corrosion	High temperature, voltage and humidity
	Oxidation	High temperature
Operation Induced	Electromigration	High temperature and current
	Hot carriers	Low temperature and voltage
	Latch-up	Voltage and current
	Electrostatic discharge	High current and voltage
Application Induced	Non-functionality	Limited fault coverage, EMI, EOS, ESD
	Electron trapping	β radiation
	Data upset	Transient radiation
	Lattice damage	Heavy charged particles, neutrons
	Soft errors	α particles

released from the plastic encapsulant. The most common mobile ion which become trapped in the oxide are Na^+ ions. Plastic encapsulants used for packaging generally release negative halide ions. With applied bias, the movement of mobile ions is accelerated. In the presence of humidity, the plastic package releases more negative halide ions. The presence of surface charge at or near the $Si - SiO_2$ interface affects the threshold voltage of the transistor. Surface inversion due to surface charge movement at the oxide surface also causes a loss of current gain and high junction leakage in the bipolar transistors. H-ions evolved from the plasma nitride interacting with the organic groups in carbon based spin-on-glass, used for intermetal planarization, may cause field inversion. A charge accumulation at $Si - SiO_2$ interface also may form a parasitic MOS structure, which may cause circuit malfunction.

Oxide contamination, during or after its growth, directly affects its dielectric properties, particularly its breakdown voltage. The electric field across the oxide accelerates the oxide breakdown. With the application of electric field to the oxide, charges are trapped, depending upon the contamination level and the intrinsic oxide structure. When the built-in field caused by the trapped charges exceeds a critical value, breakdown occurs. For a given stress field, if one waits long enough, oxide breakdown is observed. This suggests that the oxide breakdown is time-dependent. Other oxide related failures occur due to dipole polarization or charge storage effects in the oxide due to the cycling of applied bias.

To form ohmic contacts, aluminum is alloyed with silicon below the Al-Si eutectic temperature of 577°C. During the 400–450°C alloying step, solid-state diffusion of silicon into aluminum occurs, and the Al-Si interface moves down into the silicon. An additional 1–2% silicon in the deposited aluminum helps in reducing the dissolution of silicon from the junction into aluminum. Nevertheless, overalloying in shallow junctions, required in VLSI circuits, leads to junction shorting or spiking. In advanced VLSI fabrication processes, barrier metals such as Ti/TiN are used to minimize the junction spiking during the contact alloying. However, diffusion through imperfections in the barrier layer still may cause reliability problems.

Another major cause of metal interconnection failure is the formation of microcracks. This occurs when the metal coverage on steep oxide steps (such as high-aspect-ratio contacts and vias in submicron geometries) becomes poor. Steeper steps lead to thinner metal deposits, which have a higher probability of failure than normal deposits, both under temperature and current stresses as well as stresses in the surrounding thin films such as the passivation layer. Semiconductor vendors ensure a minimum metal step coverage by continuously monitoring it, using scanning electron microscopes.

Etchant residues due to insufficient cleaning and surface passivation after the metal etching, halides released by the aging plastic encapsulation, or the presence of phosphorus in oxide in contact with metal lines may lead to the chemical corrosion of the metal lines. In the presence of humidity, electrolytic corrosion also may occur. Aluminum and its alloy films generally corrode in the presence of OH^- ions. Semiconductor manufacturers use low phosphorus oxides and elaborate post-etch cleaning and passivation techniques to avoid metal corrosion. Other reasons of metal failures include poor adhesion to oxide, poor bondability, and poor ohmic contacts.

12.4.2 Assembly Related Failures

Assembly related failures can be divided into three general categories: (1) *chip mounting related*, (2) *wire bonding related*, and (3) *packaging material related*.

Poor process control, improper handling, contaminated solder balls, and interfacial stresses may cause voids and cracks in bonding. These failures are enhanced at higher temperatures. Mismatch of thermal coefficient of silicon, epoxy (die-attach material), and plastic or ceramic packaging may cause mechanical stresses, leading to die-attach failure or die cracking during temperature cycling. Epoxy-mounted chips are especially vulnerable because the thermal coefficient of expansion of many epoxies exceeds by an order of magnitude that of the silicon or the packaging substrate. Moreover, if the epoxy is not properly cured, its thermal properties and expansion characteristics may not be controllable. Thermal mismatch may create hot spots, causing thermal runaway. Devices assembled by the "solder bump" technique could also be affected by thermal stresses.

Chips mounted with eutectic bonding generally have a better match in thermal coefficients. However, poor control in the composition of eutectic material may cause changes in spatial composition and segregation of alloy material during power cycling. This may lead to poor thermal conductivity and mechanical bonding. In some cases, when boron-doped eutectic preforms are used to provide P^+ doping of the backside of p-silicon, spatial segregation of the bonding material may lead to poor backside contact and parametric change. Oxidation of eutectic material at elevated temperatures may lead to poor heat conduction and degradation of the backside contact. Poor heat conduction results in higher junction temperature, and thus a possible failure.

Chips are usually bonded by using gold or aluminum wires. Gold wires are bonded by thermocompression to the aluminum alloy bonding pads, whereas aluminum wires are bonded ultrasonically. Solder bumps and beam-lead techniques are also used in some packages. In the case of gold wire bonds, electrochemical potential between aluminum and gold leads to the formation of intermetallic compounds. The formation of intermetallic phase $AuAl_2$ is popularly known as "purple plague" due to its color. Higher temperature or bias stress enhances the formation of purple plague. This gold-aluminum interaction causes degradation and failure accompanied by void formation.

In the case of aluminum wires, control of wire diameter, tensile strength, and ductility are important factors. Generally, aluminum with 1% silicon is used to add hardness to the wire. Poor process and material control and contamination may lead to cracks at the heel of the bond, overbonding, or underbonding. Stress relaxation at over- or underbonded pads during high temperature stress may lead to bond failures. In some cases, bonding stresses may cause bond failures during mechanical vibration or high gravitational pull. Temperature cycling may cause fatigue stress related failures if the bond is weak due to contamination.

Metal cans and ceramic and plastic packages are used for ICs. The hermeticity of metal cans and ceramic packages depends on glass-to-metal and side-brazed seals, which may be affected by the mismatch of thermal expansion coefficients of the materials. A poorly controlled sealing or thermal cycling may lead to failure.

In the case of plastic packages, environmental factors such as humidity and temperature may cause reliability problems. The aging of plastic material releases chemicals that may affect the chip performance. Water vapor may also diffuse through the plastic package, which, if penetrated through the passivation layer, may cause electrochemical corrosion of the metallization. Any moisture trapped within the package or released from the packaging material may condense at low temperatures. This may lead to increased leakage and corrosion.

Lead finish and cleanliness of the package leads may affect the solderability of an IC in a system board. Improper storage of finished chips in a hot, humid environment may cause oxidation of the leads. An oxidized or corroded IC pin may cause costly field failures of a computer system. All these failure mechanisms are accelerated by environmental factors such as temperature and humidity. These mechanisms are also affected by temperature cycling and electrical bias. With proper process and material control, highly reliable packaging is routinely achieved.

12.4.3 Operation Induced Failures

Many failures can be directly traced to the circuit design. Some failures such as timing problems, noise generation during high speed switching, and drift in switching thresholds, may occur due to circuit design. These failures are usually guard-banded during design as well as the final test. There is another category of faults, called *operation induced faults*. These faults not only change the circuit parameters, but also result in a change in logic value [6]. These faults primarily are related to the transistor structure, layout, and circuit operation. These failure mechanisms are hot-carrier-related parametric shift, latch-up induced failure, electromigration, and electrostatic discharge.

Hot Carrier Effects

As the transistor size is scaled down, the supply voltage is not reduced in the same proportion. As a result, hot-carrier-induced reliability and performance degradation become a major reliability concern. Hot carriers are defined as the charge carriers having energy greater than the $Si - SiO_2$ energy barrier.

The effect of hot carriers in an n-MOS transistor is shown in Figure 12.4, where electrons moving from the source to the drain are accelerated by a high field at the drain. Due to impact ionization, high energy electron-hole pairs are generated. These hot electrons are injected into the gate oxide and trapped there. This causes a drift in threshold voltage or degradation of transistor transconductance. The majority of hot holes are collected as substrate current. If the substrate current is high enough, and the substrate resistivity also is high, the source-substrate

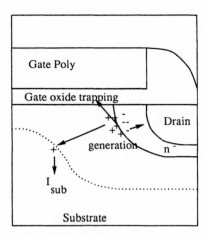

Figure 12.4 Hot carrier effect in an n-MOS transistor.

junction may be forward-biased, which results in transistor snap-back or latch-up. High substrate current also may overload the substrate bias generator.

The hot carrier effect is observed in both n-MOS and p-MOS transistors. The effect is more severe in n-MOS due to smaller p-Si/SiO$_2$ barrier height as compared to n-Si/SiO$_2$ barrier in p-MOS transistor. The hot carriers may originate either from the surface channel current (channel hot carriers), or from the drain avalanche (drain avalanche or substrate hot carriers). Channel hot carriers cause a degradation in channel mobility, and hence a degradation in the transistor transconductance due to interface state generation. If V_{DS} is sufficiently large, reducing V_{GS} will raise the channel electric field up to the point that avalanche multiplication increases the number of hot carriers. It happens when the drain bias is much more than the gate bias (maximum effect occurs when the drain bias is about twice the gate bias). The substrate hot carriers cause a shift in threshold voltage due to charge trapping, which occurs when the gate bias is applied while the source and the drain junctions are in avalanche breakdown. In general, degradation in both transistor transconductance and threshold is observed [3]. The substrate hot carrier effect becomes more severe at high temperature, whereas the channel hot carrier effect increases at lower temperature [4].

Latch-up

Latch-up is defined as a state of high excess current accompanied by the switching of nearby transistors. Latch-up generally disrupts the functional capability of the circuit and sometimes causes permanent damage to the circuit. In the literature, three regenerative mechanisms are defined as latch-up:

1. Bipolar transistors sustaining voltage breakdown caused by the current injection from the emitter into the base, thereby lowering the reverse voltage needed for avalanche breakdown in collector-base junction;
2. Secondary breakdown, which results from thermal runaway current crowding as hot spots formed at defects;
3. Four-layer (*pnpn* or *npnp*) SCR structures, i.e., regenerative switching, which shows a positive feedback between the coupled *pnp* and *npn* parasitic bipolar transistors, and allows a very large current to flow.

Due to parasitic *npn* and *pnp* bipolar transistors, a *pnpn* or *npnp* type of SCR structure is inherent to CMOS. In Figure 12.5, parasitic bipolar transistors are shown in a CMOS inverter circuit. The resulting SCR structure is shown in Figure 12.6. A criterion for latch-up of this parasitic SCR is [5]:

$$\alpha_{ns} + \alpha_{ps} \geq 1 \tag{12.8}$$

where

$$\alpha_{ns} = \frac{R_s \cdot \alpha_n}{R_s + r_{en} + R_{en} + R_{bn} \cdot (1 - \alpha_n)}$$

and

$$\alpha_{ps} = \frac{R_w \cdot \alpha_p}{R_w + r_{ep} + R_{ep} + R_{bp} \cdot (1 - \alpha_p)}$$

where α_n and α_p are common base current gains of parasitic NPN and PNP transistors; R_w, R_s, R_{en}, R_{ep}, and R_{bp} are the lumped equivalent resistances, as shown in Figure 12.6; r_{en} and r_{ep} are small-signal emitter resistances, defined as dV_{be}/dI_e.

Latch-up can be minimized by reducing the gain of bipolar transistors or decoupling the *npn* and *pnp* transistors. Early techniques to reduce the bipolar gain include gold doping or neutron irradiation, lowering the carrier lifetime by internal gettering, and use of Schottky diodes. However, these techniques proved to be either impractical or led to adverse side effects in transistor characteristics. More popular processing techniques to reduce latch-up include the use of a highly doped substrate beneath a lightly doped epitaxial layer to shunt the lateral parasitic bipolar transistor, use of a retrograde well to reduce the well's sheet resistance, and use of trench isolation to eliminate the lateral current flow in the SCR structure.

Several design and process techniques have been developed to ensure a latch-up-free operation. Majority carrier guard ring, multiple well, and substrate contacts

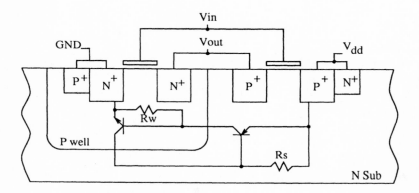

Figure 12.5 Parasitic bipolar transistors in *p*-well CMOS.

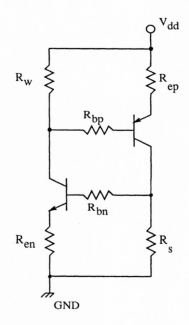

Figure 12.6 Equivalent circuit of SCR during latch-up in CMOS.

are used to minimize the ohmic drop from the parasitic collector current. Dummy wells are used to precollect minority carriers injected into the substrate before they reach an active well containing the base of a vertical bipolar transistor. For transistors operating with a grounded source, butted source contacts reduce the bypass resistance for the source's parasitic emitter behavior. A design technique to suppress latch-up in n-well bulk CMOS circuits uses a low impedance substrate bias generator. This substrate bias generator has two effects. First, by biasing the p type substrate at a negative voltage (about -3 V), the parasitic diodes cannot be forward-biased unless the undershoot exceeds the back bias. Second, the bias generator is usually enough to stop sufficient trigger from being generated.

Electromigration

A well documented metallization failure is electromigration, which is caused by the mass transport of metal atoms by the momentum exchange with conducting electrons. The movement of metal atoms is thought to occur along the interfaces between metal grains. At high current density and high temperature, metal atoms are swept by the impact of current flow, causing a hillock formation on the positive end of the conductor and voids in the opposite direction. Electromigration mainly depends on the metal additives, metal grain size, line widths, passivation layer, temperature, and current gradient. For pure aluminum lines, a current density of about 10^5 A/cm^2 induces electromigration at room temperature. As other impurities are added, minimum current density increases. For example, for Al-4% Cu alloy, electromigration occurs at about 10^6 A/cm^2. In general, mean-time-between-failure due to electromigration is given by:

$$\text{MTBF} = \frac{A}{CJ^2} \cdot \exp\left(\frac{E_a}{kT}\right) \tag{12.9}$$

where A is the cross-sectional area of the conductor, C is a constant, J is current density, and E_a is the activation energy required for electromigration. For advanced processes with metal line widths of about 1 μm, aluminum lines are sandwiched between refractory metals to improve current-carrying capacity.

In VLSI circuits with submicron contact size, contact temperature is increased due to high current densities. This may lead to electromigration of silicon in aluminum from negatively biased contacts to positively biased contacts. As a result, negative contacts may become short-circuited, and silicon precipitates increasing the contact resistance in the positive contacts. This electromigration becomes significant when a dc current exceeds about 0.2 mA/μm of the perimeter around the contact holes. Contact plugs made up of tungsten or deposited polysilicon help to reduce electromigration in the contacts and vias.

Electrostatic Discharge

Extensive use of VLSI circuits in high noise environments requires reliable *electrostatic discharge* (ESD) protection for the I/O pads. Because of the extremely high input impedance and relatively low breakdown voltage, MOS transistors have always suffered from ESD failures. Advanced processes using thin-film dielectrics, shallow junctions, lightly doped drains, and silicides make this problem worse. At submicron geometries, even bipolar transistors become sensitive to ESD because of the shallow junctions.

The ESD failure mechanism is an electrothermomigration induced by nonuniform current density and heating. This results in thermal damage to oxide, melting of metal and polysilicon conductors, and melt filament formations. Because of the complexity of structural geometries, transient microplasma, and hot spot formation, accurate transient thermal analysis is difficult. However, simple models suggest that the $J \cdot E$ product (current density × electric field) is directly related to the electrothermomigration. Location of maximum $J \cdot E$ product in a design indicates the possible point of ESD failure. A model is developed in [7] to explain the observed ESD degradation with shallow junctions, LDD junctions, silicide diffusions, shorter contact-to-gate spacing, and smaller diffusion overlap of contacts. There is a direct relationship between the ESD hardness and the snapback voltage of n-MOS protection circuit. Snapback voltage is dependent on the current nonuniformities of the transistor and are caused by the charge trapping in the oxide [8]. For uniform current flow, the ESD protection is proportional to the transistor width. However, a nonuniform current flow may reduce the effective width of the ESD protection transistor.

Most of the ESD improvement techniques are based on reducing current nonuniformities, reducing the peak of $J \cdot E$ product, and snapback voltage. To improve the current uniformity, bent corners are avoided and a waffle type of transistor layout is used. An increase in snapback resistance also helps to improve the current uniformity. Wide transistors, large contact-to-gate space, deep junctions, and increased diffusion overlap of contacts help in reducing the $J \cdot E$ product. In advanced processes, some of these solutions in the I/O area may not be compatible with the rest of the circuit. Snapback voltage of the protection transistors may be reduced by using minimum effective channel length, a small positive gate bias with respect to the source, and a small forward bias at the source-substrate junction. In general, a secondary transistor is used in conjunction with the primary ESD protection transistor. For input protection, a thick-gate oxide FET is used in parallel with a thin-gate oxide FET as shown in Figure 12.7. For output buffers,

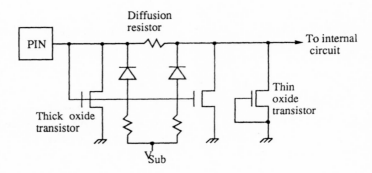

Figure 12.7 A typical ESD protection circuit.

the output transistors are laid out in conjunction with a thick-gate oxide FET to provide ESD protection. Recently, low voltage triggering SCRs have been used to provide ESD protection. Just designing I/O ports for ESD protection is not enough. Layout of the power-supply and the ground busses inside the chip should also be considered to ensure that a weak parasitic transistor would not lower the overall ESD reliability.

12.4.4 Application Induced Failures

Many times, two equivalent circuits made by two different vendors, or by the same vendor in two different batches, behave differently in a given application. The same IC, which otherwise functions properly and equivalently in many applications, may fail early or not work at all in some applications. Obviously, the IC which fails early has some parameter that is more sensitive in a specific application than others. One needs a perspective on the failures that are induced by the specific way in which an IC has been used.

Some of the most common application-induced failures are related to the drift in design margins or different guard bands used by different vendors. This includes the design margin for system-generated noise in the ground and power supply bus, capacitive loading of the I/Os, timing skews, and I/O level swings. Because of the statistical variations in chip and board manufacturing, a particular batch may show different parameters.

Incomplete testing of all possible functional faults in microprocessors, micro-controllers, random logic, ASICs, and memories may also cause application-induced failures. Incomplete testing of all possible functional faults in a circuit is directly related to the incoming quality factor. In memories, some bits may have varying degrees of data retention time, which may not be fully tested. In highly reliable fault-tolerant designs, these possibilities are always considered. However, in many commercial designs, they lead to application-induced faults. Proper guard-banding of the circuit can minimize such failures.

Another application-induced failure is related to *electrical overstress* (EOS) and environmental factors, i.e., radiation and *electromagnetic interference* (EMI). Overstress and wear-out failures of an IC due to voltage, current, temperature, and other operating parameters may be avoided by a good understanding of the IC life characteristics. Electromagnetic radiation produced by output switching may interfere with the system functionality, or with the on-chip clocks. EMI effects are best solved by implementing Faraday shields at the system level, package level, and chip level.

One particularly important environmental factor for military and space applications is the ionization radiation. Single-event upset caused by low-level radiation is also important to commercial memories, particularly for DRAMs. Changes in transistor characteristics depend on the type of radiation. For military and space applications, total radiation dose exposure, transient ionizing radiation, photon radiation, heavily charged particles, neutron fluence, and single-event upsets are of concern. In a MOS circuit, exposure to ionizing radiation over a period of time (total dose) creates dipoles in the oxide and silicon. This causes a shift in threshold voltage, degradation in carrier mobility, surface inversion, and increased leakage current. Special radiation-hardened processes are used to improve the total dose hardness. Silicon-on-insulator or sapphire (SOI/SOS) provides a good hardness against radiation. Thinner oxides used in advanced processes also improve total dose hardness.

Transient ionizing and photon radiation (gamma particles) from a nuclear event generate electron-hole pairs and photocurrents. This may change logic states and cause circuit ringing. Particularly vulnerable are bistable circuits such as latches, flip-flops, and memory cells. If the photocurrent is large enough, it may also cause junction, interconnection, and wire bond failures. Heavily charged particles and neutron radiation may damage the crystalline lattice structure, affecting carrier mobility, minority carrier lifetime, and junction leakage. Bipolar transistors are especially affected by neutron radiation. In general, MOS circuits do not exhibit serious performance degradation due to neutron radiation until the dose exceeds 10^{15} neutrons/cm^2. This immunity exists mainly because MOSFETs are majority carrier devices.

A single-event upset or soft error occurs when an ionizing particle hits a logical node in a circuit. Particles causing soft errors include protons or heavier ions in cosmic rays, α particles from radioactive contamination in packaging material, and other radioactive sources.

The critical dose required for a radiation induced error depends on circuit design as well as on manufacturing process. For example, a six transistor CMOS SRAM cell is less prone to soft errors than a four-transistor SRAM cell. As the circuit dimensions shrink, total dose tolerance increases due to reduced oxide volume, which, in turn, reduces the cross section of radiation absorption. Use of guard rings, epitaxial material, good stoichiometric thin oxide, and low temperature processing also improve radiation hardness. Use of thick polyimide or silicone coating and low-radioactive-content packaging material help to reduce the soft errors. Silicon-on-insulator or sapphire, gallium arsenide, and diamond-based circuits have much higher total radiation hardness than conventional bulk silicon circuits.

12.5 RELIABILITY TEST METHODS

The purpose of reliability testing is to ensure that the circuit will perform its stated functions under given operating conditions for at least a specific period of lifetime. The screening process selects chips with superior reliability and rejects those which potentially will fail early during operation. The number and type of reliability tests used in a particular application are determined by the economic benefits of the reduced component failure rate and cost of screening. Military Standard 883 [9], for example, specifies a stringent reliability test for Class S circuits used for space, where maintenance and replacement are impossible, as compared to Class B circuits used for avionics and other military applications. Commercial ICs have even less comprehensive reliability tests.

Determination of reliability by testing under actual operating conditions for the entire lifetime is not feasible. Therefore, reliability testing consists of several tests designed to excite expected failure mechanisms. Further, reliability tests are spread throughout the entire manufacturing process to achieve continuous screening for various failure mechanisms. Table 12.2 shows various reliability tests performed at the transistor, assembly, circuit, and batch levels to ensure overall reliability. Generally, a statistical sample size is chosen for these tests to meet the desired reliability goal. The detailed descriptions of these test methods are given in Military Standard 883. Test reference numbers in Military Standard 883 are also given in Table 12.2.

After manufacturing and various reliability tests, ICs are put to quality conformance tests. The extent of these conformance tests depends on the customer's requirements. These are also done on a randomly chosen sample, except for electrical functionality tests. Examples of military Class S and Class B quality conformance tests are given in Table 12.3. Commercial chips are generally subjected to a subset of Class B tests.

Finished chips are subjected to further reliability tests designed to screen out "infant" mortality. Table 12.4 lists the early failure screening tests typically specified by the military. A sample of the chips is subjected to wear-out life tests specified in Table 12.5 to determine the typical lifetime of the circuit.

Table 12.2
Reliability test methods for ICs.

Test Level	Parameter Tested	Military Standard 883 Method Number
Transistor	Dielectric breakdown	5010
	Hot carrier effect	5010
	Electromigration	5010
	Junction temperature	1012
	Metallization quality	2018
	Passivation integrity	2021
	Radiation effect	1017, 1019, 1020, 1021, 1023
	Memories endurance	1033
	Mobil ion contamination	2500 (Mil-Std 977)
Assembly	Die attach integrity	2012, 2019, 2017, 2030
	Wire bond strength	2011, 2023, 2028, 2031
	Seal integrity	1002, 1014, 2012, 2024
	Moisture content	1002, 1003, 1013, 1018
	Thermal integrity	1010, 1011, 1012
	Corrosion resistance	1009, 1031
	Solvent resistance	2015
	Package induced soft errors	1032
	High altitude effects	1001
	Vibration/mechanical shock	2001, 2002, 2005, 2006, 2007, 2026
	Lead integrity/solderability	2003, 2004, 2025
Circuit	ac, dc, functionality	3001–3011, 3014, 3020–3022, 4001–4007
	Noise margin	3013
	Crosstalk, latch-up	3017, 3018
	ESD, EMI, power dissipation	3015
Batch	Wafer lot parametrics	5001, 5002, 5007
	Packaging visual defects	2013, 2014, 2016, 2017, 2020
	Early failure	5004, 5005, 5008, 5010
	Wear-out	5006, 5003

12.6 ACCELERATED RELIABILITY TESTING

For a proper accelerated reliability test, circuit failure mechanisms and the acceleration factors must be well understood. A particular stress generally accelerates more than one failure mode. To ensure that all major failure modes are covered, different accelerated tests are required. Also, one should realize that accelerated reliability testing is not a substitute for field testing. At least one field test on a large number of chips for a long period of time under expected operating conditions is always desirable.

Table 12.3
Quality conformance tests for ICs.

Group	Test	Class S	Class B
(A) Electrical Tests	Static test at 25°C, min. and max. temp.	x	x
	Dynamic test at 25°C, min. and max. temp.	x	x
	Functional test at 25°C, min. and max. temp.	x	x
	Switching test at 25°C, min. and max. temp.	x	x
(B) Assembly Tests	Physical dimensions	x	x
	Particle impact noise detection	x	—
	Resistance to solvents	x	x
	Internal visual and mechanical inspection	x	x
	Destructive wire bond strength	x	x
	Die shear strength	x	x
	Solderability	x	x
	Fine and gross seal leakage	x	x
	ESD	x	x
(C) Package Stress Tests	External visual inspection	x	x
	Temperature cycling, thermal shock	x	x
	Mechanical shock, constant mech. acceleration	x	x
	Fine and gross seal test	x	x
	Radiograph of the package	x	—
	Visual examination	x	x
	Electrical test	x	x
	1000 hr, 125°C life test	x	x
(D) Environmental Tests	Internal moisture content test	x	x
	Moisture resistance test	x	x
	Salt corrosion resistance test	x	x
(E) Radiation Hardness Tests	Neutron radiation	x	—
	Steady state total dose irradiation	x	—

Major stress factors to accelerate reliability testing are temperature, voltage, and current for circuit-level failures. Humidity, temperature cycling, power cycling, vibration, and shock are the acceleration factors for assembly and packaging related failures.

Both *constant-stress* and *step-stress* techniques are used for accelerated reliability testing. If the failure rate is extremely small for a particular failure mechanism, step-stress techniques considerably reduce the test time. For example, electromigration testing can be done by increasing the current in steps. Similarly,

Table 12.4
Early failure screening of ICs.

Reliability Screen	Military Standard 883 Method	Class S	Class B
Wafer lot acceptance	5007	x	—
Nondestructive bond pull	2023	x	—
Internal visual inspection	2010	x	x
Stabilization bake	1008	x	x
Temperature cycling	1010	x	x
Constant mechanical acceleration	2001	x	x
Visual inspection		x	x
Particle impact noise detection	2020	x	—
Device marking, serialization		x	—
Pre-burn-in test	as appropriate	x	x
Burn-in test	1015	x	x
Interim post burn-in test	as appropriate	x	—
Reverse bias burn-in	1015	x	—
Interim post burn-in test	as appropriate	x	x
Percent defect calculation		x	x
Final electrical test	5005, 5008, 5010	x	x
Fine and gross seal test	1014	x	x
Package radiograph	2012	x	—
Quality conformance test	5005, 5008, 5010	x	x
External visual inspection	2009	x	x

Table 12.5
Sample size for wear-out life tests of ICs.

Test	Sample Size
Thermal evaluation	5
Extended thermal shock	10
Step-stress mechanical shock	10
Step-stress constant acceleration	10
Step-stress operational life test	10
Constant high stress operation life test	10
Step-stress storage life test	10

dielectric breakdown or electrostatic discharge testing can be done by increasing the voltage in steps. In many cases, two or more stress factors are simultaneously applied to accelerate the testing further. Temperature and voltage stresses are generally combined to accelerate the circuit life test. Similarly, temperature and humidity stress are combined to accelerate the plastic package failures.

12.6.1 Temperature Acceleration

Many failures occur because of various chemical or physical processes that may be accelerated by the temperature. For a chemical or physical process, the reaction rate, R, can be expressed as

$$R = R_0 \exp \left(\frac{-E_a}{kT} \right) \tag{12.10}$$

where R_0 is the reaction rate constant, which generally is a function of time, current density, voltage, and temperature. E_a is the activation energy for the chemical or physical process, k is Boltzmann's constant, and T is the absolute temperature.

For a reaction in which R_0 is not dependent on time and temperature, the above equation is called the Arrhenius equation. The acceleration factor due to temperature is given by

$$A_f = \exp \left[\left(\frac{E_a}{k} \right) \cdot \left(\frac{1}{T_1} - \frac{1}{T_2} \right) \right] \tag{12.11}$$

where T_1 and T_2 are two operating temperatures such that $T_2 > T_1$. The activation energy is determined by testing several samples at different temperatures for the lifetime. Each test provides a statistical distribution (usually log-normal) for the lifetime. Median life is the least affected parameter by sampling variation, and therefore is of interest. It can be determined by the Arrhenius plot, graph of ln(median life) versus $1/T$. In this calculation, it is important to ensure that, at a given stress temperature, (1) the failure rate is not a function of stress time; (2) major failure mechanisms do not change during the stress period; (3) the shape of life distribution ($\sigma = \ln t_x/t_y$) is same over the temperature range, where t_x and t_y are time to $x\%$ and $y\%$ failures.

Also important is to account for a rise in chip temperature due to the power dissipation. For a packaged chip, the junction temperature is given by

$$T_j = T_a + P \cdot \theta_{ja} \tag{12.12}$$

where T_j is the junction temperature of the IC, T_a is the ambient temperature, P is power dissipation, and θ_{ja} is junction to ambient thermal resistance. For example, for a circuit dissipating 0.25 W, with $\theta_{ja} = 40°C/W$, junction temperature will be about 10°C higher than the ambient.

In general, several failure mechanisms can be excited at a given temperature. As shown in Figure 12.8, the failure distribution may be bimodal or multimodal [10]. In such cases, the main failure rate is given by

$$R_T(t) = R_s(t) \cdot n_s + R_m(t) \cdot n_m \qquad (12.13)$$

where $R_T(t)$ is the total failure rate at temperature T and time t, $R_s(t)$ is the secondary failure rate at time t for n_s samples, and $R_m(t)$ is the major failure rate at time t for n_m samples. Table 12.6, based upon [11], summarizes the activation energies for various failure mechanisms.

Example 2. Let us assume a 10°C increase in junction temperature due to power dissipation and an activation energy of 0.7 eV, the failure at 125°C ambient stress temperature will be accelerated by about 279 times as compared to 35°C application temperature. Therefore, 1000 hours of stress at 125°C is equivalent to about 32 years of operation at 35°C. Further, if two chips fail out of 1000, tested during 1000 hours of 125°C stress, χ^2 in Equation (12.5) for a 60% upper confidence limit will be 6.21. This means a failure rate of 3105 FIT at 125°C and a failure rate of 11 FIT at 35°C with a 60% confidence limit.

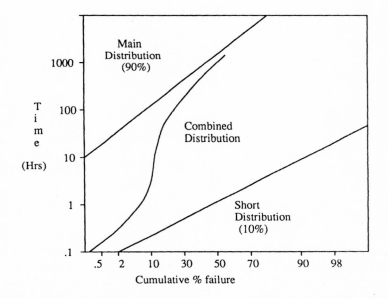

Figure 12.8 Bimodal failure distribution. (From [10], © 1971 IEEE. Reprinted with permission.)

Table 12.6
Activation energies for various failure modes in CMOS.

Failure Type	Failure Mode	Activation Energy (eV)
Oxide and oxide interface	Dielectric (oxide) breakdown	0.3
	Silicon defects	0.3
	Surface charge accumulation	1.0
	Slow charge trapping	1.0
	Microcracks	1.3
	Contamination	1.4
	Hot carrier trapping	-0.06
Interconnect	Electromigration of Al	0.5
	Corrosion	0.8
	Electromigration of Si in Al	0.9
	Contact degradation	1.8
Assembly	Plastic chemistry	1.0
	Polarization	1.0
	Al-Au intermetallic growth	1.02

Note that the activation energy for some failures may be dependent on the operational history. For example, in the case of EEPROMs, a decrease in activation energy for data retention failure with the number of write or erase cycles is common. For n-MOS EEPROMs, data retention activation energy is reported to decrease from about 0.67 eV for 10 write or erase cycles to about 0.25 eV for 10^5 write or erase cycles [12]. Also, note that the failure activation energy is not necessarily a positive number. For example, substrate current due to hot electrons decreases at higher temperature [4] because the injected hot electrons are released from the gate oxide at higher temperature. Therefore, very little degradation in the threshold voltage and transconductance due to hot electron injection is observed at high temperatures.

12.6.2 Current Acceleration

Metallization-related failures are accelerated by current as well as temperature. The median time to failure of interconnections (t_{50}), caused by electromigration due to dc current, is given by Equation (12.9), rewritten as [13]:

$$\frac{w \cdot t}{t_{50}} = RJ^N \exp\left(\frac{-E_a}{kT}\right) \tag{12.14}$$

where w is metal line width, t is metal line thickness, J is dc current density, E_a is the electromigration thermal activation energy (0.5 eV), k is Boltzmann's constant, and N is the current density factor (although experimentally determined values for N are in the range of 1 to 7, reference [13] suggests $N = 2$). R is the reaction rate constant, given as

$$R = A\rho\left(\frac{l}{\nu}\right)\sigma_c \qquad (12.15)$$

where A is a constant dependent on metal composition, ρ is volume resistivity, l is the electron mean free path, ν is average electron velocity, and σ_c is scattering cross section. In the case of electromigration occurring in the metal contacts to the silicon diffusion areas, contact size and junction depth also affect the reaction rate. The activation energy for silicon contact electromigration is also different, as shown in Table 12.6.

Electromigration can be accelerated by increasing either current or temperature. The trade-off between current and temperature acceleration is optimized for an acceptable stress time. For aluminum interconnections, at temperatures above 350°C, the failure mechanism changes from grain boundary electromigration to bulk migration. At very high current densities, the Joule heating raises the interconnection temperature, which must be taken into account. Note that the maximum current through an interconnection is limited by the circuit design and cannot be independently varied. Also, the current flowing in an IC interconnection is not always a fixed dc, but usually is a pulse current. The median time to failure increases as a function of duty cycle of this pulse current. Due to these factors, a good current acceleration test for ICs has not yet been developed. At present, circuits are designed such that the current density in the interconnections does not exceed the allowable current density for the specified lifetime. To characterize the electromigration failures of interconnections, special wafer-level test structures are used for current accelerated stress tests [14]. Using special test structures, the acceleration factor due to current density at various ambient temperatures is shown in Figure 12.9.

12.6.3 Voltage Acceleration

High voltage or electric field accelerates oxide breakdown related failures. The MTBF of oxide due to time-dependent dielectric breakdown is given by

$$t_{50} = B \cdot \exp\left(\frac{E_a}{kT}\right) \cdot \exp[\gamma(T)\, S] \qquad (12.16)$$

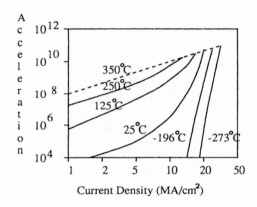

Figure 12.9 Electromigration failure as a function of current density. (From [14], © 1985 IEEE. Reprinted with permission.)

where B is a constant, and $\gamma(T)$ is a temperature-dependent field acceleration factor, which is given by

$$\gamma(T) = -4.5 + 0.27 \text{ eV/kT, in decades/MV/cm} \tag{12.17}$$

For a given oxide, the field acceleration at room temperature is about 6 decades/MV/cm, whereas it reduces to about 2 decades/MV/cm at about 150°C. For 20-nm thick-gate oxide MOSFET, operating at 125°C, there is about $30\times$ acceleration at 5.5 V bias as compared to 5 V bias, whereas, for the same oxide under the same voltage acceleration at room temperature, the oxide failure accelerates by about $129\times$.

E_a in Equation (12.16) is the activation energy for the oxide failure. In the case of MOS structure, it is the change in enthalpy required to initiate the filament growth of gate material at oxide breakdown. For polysilicon gates, E_a is given by

$$E_a \text{ (in eV)} = -0.7 + 0.35 \cdot S; \text{ with } S = E_B - E_S, \text{ in MV/cm} \tag{12.18}$$

where E_B is the oxide breakdown field (on the order of 10 MV/cm), and E_S is the stress field in the units of MV/cm. For example, for 10-nm thick oxide, with 5.5 V stress, the activation energy is about 0.9 eV (assuming $E_B = 10 \text{ MV/cm}$), whereas, for 20-nm thick oxide with the same stress, the activation energy is about 1.8 eV. The experimental data show that for low stress field (≤ 5 MV/cm), the activation energy is constant at about 0.75 eV. For higher field (≥ 6 MV/cm), the activation energy is inversely proportional to the stress field. Note that the standard deviation of log-normal failure distribution is not constant for voltage acceleration [15], and therefore modeling is not statistically unambiguous. However, Equation (12.18) explains a wide range of experimental observations.

In most ICs, the operating voltage cannot be increased beyond a certain value. Also, the effect of field acceleration reduces at higher temperature. Therefore, the oxide failures are generally studied by using field acceleration at room temperature on wafer-level MOS test structures. This wafer-level reliability test can be done by using either a constant voltage stress or constant current stress. In the case of constant current stress, the electrons trapped in the gate oxide cause the gate voltage to increase with time until the oxide breakdown occurs. The charge to breakdown (Q_{BD}) in constant current stress testing is related to time to breakdown in constant voltage stress test.

12.6.4 Temperature-Humidity Acceleration

Many failures, such as electrochemical corrosion, mobile ion contamination, and packaging-induced failures, can be accelerated by a combination of temperature and humidity stress. The MTBF of an IC with aluminum metallization and plastic package is given as [16]:

$$t_{50} \, \alpha \, (\%\text{RH})^{-2.66} \exp\left(\frac{0.79}{kT}\right) \qquad (12.19)$$

where %RH is percent relative humidity. Moisture resistance tests using temperature-humidity chambers are performed at 85°C, 85% relative humidity (called 85/85 test). With improved reliability of plastic packaging, moisture tests at higher temperature are becoming popular. Many manufacturers use a *highly accelerated stress test* (HAST) at 120–150°C, 85%RH. Note that if the chip temperature is allowed to rise above the ambient temperature due to power dissipation, the relative humidity at the chip will be less than the ambient relative humidity. Therefore, the acceleration will be offset by the decrease in humidity.

Moisture resistance testing is also performed using a pressure-cooker, in which the chips are subjected to 120°C, 15 PSIG, and 100% RH. Since the moisture-induced failure is dependent on the quality of passivation, most manufacturers perform a wafer-level passivation integrity test. In this test, the wafers are immersed in a metal etching solution. Metal corrosion caused by the metal etching solution penetrating the passivation layer through pinholes or cracks determines the quality of passivation.

To ensure moisture resistance, Military Standard 883, method 1013, specifies a maximum moisture content within the hermetic package as well as the package leak rate. The package moisture content is measured in terms of the dew point, which must be less than $-65°C$. The package leak rate is measured by using a helium leak check or radioactive tracer gas (Kryton-85) for fine leaks and fluorocarbons for gross leaks (Military Standard 883C, method 1014). For some specialized applications, where chips may be exposed to salty atmosphere, failures due to corrosion of a lead frame are also tested. This is done by subjecting chips to a salt deposition chamber at 35°C with salt deposition rate of 10–50 $g/m^2/day$ (Military Standard 883C, method 1009).

12.6.5 Vibration and Shock Acceleration

Most chips are subjected to mechanical acceleration, vibration, and shock during transport or system application. Therefore, mechanical acceleration, vibration, and shock testing are important. The basic test is specified in Military Standard 883C, method 2001. According to Military Standard 883 specifications, mechanical acceleration tests are performed using a constant acceleration of 5000–125,000 g for one minute in a centrifuge machine. For shock testing, chips are subjected to five shock pulses of 500–30,000 g, with a pulse duration between 1 to 0.12 ms at 20–20,000 Hz vibration frequency with peak acceleration of 20–70 g (Military Standard 883C, methods 2005, 2006, 2007, and 2026).

12.6.6 Temperature, Humidity, and Power Cycling

Several failure mechanisms are best accelerated by temperature, temperature-humidity, or power cycling. These tests detect failure due to thermal mismatch, fatigue-induced failures, encapsulant aging related failures, and bond failures in power devices. For temperature cycling, Military Standard 883, method 1010, specifies minimum 10 cycles of a hot-cold temperature combination of $+85°C/-55°C$ to $+300°C/-65°C$, with minimum dwell time of 10 minutes at each temperature. To test the sealing quality of the package, liquid-immersion-based temperature cycling is used (Military Standard 883, method 1002). In this cycling, the chips are immersed at 65°C and then at 0°C. Cycling of temperature-humidity accelerated testing is also performed for further testing the package reliability (Military Standard 883, method 1004).

Power cycling may cause transient temperature variation near the interconnection interface to the chip, which cannot be easily simulated by an ambient temperature acceleration test. Therefore, power cycling is also used as one of the interconnect reliability tests. In the case of EEPROMs, write or erase cycling also affects the reliability, and hence an endurance test is also performed.

Very little work has been done to relate the low stress cycling of normal operating conditions to the high stress cycling used in accelerated tests. The stress analysis of fatigue in gold beam leads shows a relationship between the stress cycles for median failures and the difference between hot and cold temperatures of the cycle [17]. In general, the acceleration factor must be determined prior to a particular stress acceleration.

12.7 BURN-IN

The failure rate generally is the highest during the early life, as illustrated in Figure 12.2. Although each manufacturer spends considerable time and effort to eliminate potential failure by using various quality and reliability checks, ensuring that the infant mortalities are screened out at the component level is important to minimize system downtime. To screen out infant mortality, burn-in is usually done. Burn-in is designed to accelerate infant mortalities to screen out potentially faulty chips. If the acceleration factor during burn-in is A_F, as compared to normal operating conditions, and the burn-in time is t_{bi}, the effective failure rate after burn-in is given by

$$\lambda_{eff}(t) = \lambda(t + t_{bi} \cdot A_F) \tag{12.20}$$

The burn-in condition generally uses temperature and voltage as accelerating factors. Temperature or power cycling also may be used as an accelerating factor. Studies have shown that infant mortality mechanisms in ICs have an activation energy of about 0.37–0.42 eV [18]. The failure mechanisms include oxide pinholes, patterning defects causing near open or short circuits, and contamination.

Military Standard 883C, method 1015, specifies the burn-in time-temperature matrix. The usual burn-in screening for ICs lasts approximately 168 to 250 hours, with stress levels of near-rated power dissipation, and temperature of 125°C. For class S ICs, a 72-hour reverse-bias burn-in at 150°C also is done. A 125°C, 168-hour burn-in condition is commonly used for commercial ICs. Most burn-in procedures use air to control the ambient temperature. For circuits with large power dissipation, the poor heat transfer characteristics of air may cause significant error, depending upon the thermal activation energy. Therefore, in some cases, liquid ambient is chosen.

Two types of burn-in are commonly used to activate a wide range of failure mechanisms: (a) *static burn-in*, in which I/O and bias pins are connected to either dc supply voltage or ground (generally done for discrete ICs); (2) *dynamic burn-in*, in which some of the inputs are clocked, (used for MOS and bipolar ICs to simulate the actual operating conditions). To minimize the total reliability testing time, ICs are often tested during the burn-in period. Another advantage of testing during burn-in is that some failure mechanisms, such as soft errors, are easily detected.

12.8 TESTING OF APPLICATION INDUCED FAILURES

No amount of accelerated testing or burn-in can provide good coverage of application-induced failures, such as radiation, ESD, and latch-up. These failure mechanisms must be tested separately. Failure mechanisms related to statistical parametric drift are tested by using guard-banded functional test programs. Military Standard 883C specifies the radiation tests required for class S ICs. For neutron irradiation, chips are exposed to a fast-burst nuclear reactor with all pins shorted (Military Standard 883C, method 1017). For total dose (measured in *rad*, radiation absorbed dose) effects, the circuits are exposed to a Cobalt-60 gamma ray source with a dose rate of 100–300 rad (Military Standard 883C, method 1019). Radiation-induced latch-up (method 1020) and dose rate threshold for device upset (methods 1021 and 1023) are tested by using a flash x-ray source or a linear accelerator. Package-induced soft errors are tested using a Thorium-232 source (method 1032).

For ESD testing, the military, IEEE, and International Electrochemical Commission (IEC) have issued different standards. For IC testing, Military Standard 883C, method 3015, is the most commonly used method [19, 20]. This standard calls for a human body model ESD pulse generated by discharging a 100 pF capacitor through a 1.5 kΩ resistor attached to the pin of the circuit-under-test. Each pin is tested with respect to all other pins and supply pins for both positive and negative ESD pulses. The human body model is considered to be a benchmark test rather than a real-world situation. Therefore, other test waveforms, such as the machine model and charged device model have also been used [8].

Depending on the board design, the board-level ESD may be quite different from the IC-level ESD sensitivity. For system-level ESD testing, IEC Standard 801-2 is often used. This is a variation of the human body model, with a 150 pF capacitor discharging through a 330 Ω resistor. An 8-mm diameter sphere is used for air discharge tests and a pointed tip is used for contact discharge tests. In commerical ESD simulations, tips are available that generate *E*-field or *H*-field pulses. Low impedance circuits tend to be sensitive to *H*-field ESD because the *H*-field induces current pulses. Conversely, high impedance circuits are affected by the *E*-field ESD pulses. To avoid damage to the circuits, sometimes an *electrical fast transient* (EFT) test is used as an indirect ESD test [21].

For CMOS ICs, the latch-up characteristic is also of importance. Most commonly, the latch-up is induced by glitches at the I/O pins, which may provide the necessary trigger current. JEDEC Standard Number 17 provides a test for this purpose [22]. In this method, a trigger current is either forced in or forced out of the pin under test, while the circuit is biased, all the untested input pins are tied to the supply voltage, and all untested output pins are open. A latch-up condition is detected by a sudden increase in the supply current. To avoid damage to the chip, a current compliance is set on the power supply. Tests for radiation-induced latch-up are specified in Military Standard 883C, method 1020.

12.9 SUMMARY

Quality and reliability testing of an IC is an important aspect of an overall IC testing program. Basic testing methods have been discussed to help establish the quality and reliability of the finished ICs. These tests may become the greater part of the total manufacturing cost. If these tests are pruned to reduce cost, customer-specified quality and reliability cannot be achieved. The quality and reliability must be built in to the design and manufacturing processes, rather than solely provided by screening and testing. This approach has been the key to several orders of magnitude improvement in the reliability of commercial ICs, such that, in many cases, commercial ICs routinely meet the strict military quality and reliability standards.

PROBLEMS

1. A circuit board designer bought 10,000 chips from a manufacturer who tests the ICs for 99% fault coverage. How many chips does the designer expect to be faulty? Assume the relationship between fault coverage and AQL as given in Section 12.1.
2. How many hours of IC testing are required to predict the failure rate with an upper confidence limit of 70% for single failures? Consider that the current IC failure rate is 5 FIT.
3. An accelerated test is performed at 100°C and 5.5 V bias. What is the gate oxide failure accelerated factor as compared to a test performed at room temperature (27°C) and 5 V bias? (Hint: use Equations (12.11) and (12.17).)

REFERENCES

1. L.R. Goldthwaite, "Failure Rate Study for the Log-normal Lifetime Model," *Proc. IEEE Symp. Rel. and Quality Control*, 1961, p. 208.
2. D.L. Denton, "Failure Rate Calculation Using a Programmable Calculator," *IEEE Trans. Rel.*, 31(4), Oct. 1982, pp. 324.
3. F.C. Hsu and S. Tam, "Relationship between MOSFET Degradation and Hot Electron Induced Interface State Generation," *IEEE Elect. Dev. Letts.*, 5(2). Feb. 1984, p. 50.
4. C. Hu, S.C. Tam, F.C. Hsu, P.K. Ko, T.Y. Chan, and K.W. Terrill, "Hot Electron Induced MOSFET Degradation—Model, Monitor, and Improvements," *IEEE Trans. Elect. Dev.*, 32(2), Feb. 1985, p. 375.
5. D.B. Estreich, "The Physics and Modeling of Latch-up and CMOS Integrated Circuits," Ph.D. Dissertation, Department of Electrical Engineering, Stanford University, Oct. 1980.
6. R. Rajsuman, A.P. Jayasumana, Y.K. Malaiya, and J. Park, "An Analysis and Testing of Operation Induced Faults in MOS VLSI," *Proc. IEEE VLSI Test Symp.*, 1990, p. 137.
7. C. Duvvury, R.N. Rountree, H.J. Stiegler, T. Polgreen, and D. Corum, "ESD Phenomena in Graded Junction Devices," *Proc. IEEE Int. Rel. Phys. Symp.*, 1989, p. 71.

8. N. Khurana, T. Maloney, and W. Yeh, "ESD on CHMOS Devices—Equivalent Circuits, Physical Models and Failure Mechanisms," *Proc. IEEE Int. Rel. Phys. Symp.*, 1985, p. 212.

9. Military Standard 883C, "Military Test Methods and Procedures for Microelectronics," Change Notice 12, National Standards Association, Gaithersburg, MD, 1990.

10. D.S. Peck, "The Analysis of Data from Accelerated Stress Tests," *Proc. IEEE Int. Rel. Phys. Symp.*, 1971, p. 69.

11. D.S. Peck, "Practical Applications of Accelerated Testing—Introduction," *Proc. IEEE Int. Rel. Phys. Symp.*, 1975, p. 253.

12. T. Ajiki, M. Sugimoto, H. Higuchi, and S. Kumada, "Temperature Accelerated Estimation of MNOS Memory Reliability," *Proc. IEEE. Int. Rel. Phys. Symp.*, 1981, p. 17.

13. J.R. Black, "Electromigration—A Brief Survey and Some Recent Results," *IEEE Trans. Elect. Dev.*, **16**, 1969, p. 338.

14. C.C. Hong and D.L. Crook, "Breakdown Energy of Metal (BEM)—A New Technique for Monitoring Metallization Reliability at Wafer Level," *Proc. IEEE Int. Rel. Phys. Symp.*, 1985, p. 108.

15. J.W. McPherson and D.A. Baglee, "Acceleration Factors for Thin Gate Oxide Stressing," *Proc. IEEE Int. Rel. Phys. Symp.*, 1985, p. 1.

16. D.S. Peck, "Comprehensive Model for Humidity Testing Correlation," *Proc. IEEE Int. Rel. Phys. Symp.*, 1986, p. 44.

17. J.L. Dais and F. Howard, "Fatigue Failures of Encapsulated Gold-Beam Lead and RAB Devices," *IEEE Trans. CHMT*, **1**, 1978, p. 158.

18. D.S. Peck, "New Concerns about Integrated Circuit Reliability," *Proc. IEEE Int. Rel. Phys. Symp.*, 1978, p. 1.

19. IEEE C62.47, "Guide on Electrostatic Discharge from Personnel and Mobile Furnishings," Draft 10, May 23, 1990, IEEE Press, New York.

20. IEC Publication 801-2, "Electromagnetic Compatibility for Industrial Process Measurement and Control Equipment: Electrostatic Discharge Requirements," IEC, American Standards Institute, New York, 1991.

21. IEC Publication 801-4, "Electromagnetic Compatibility for Industrial Process Measurement and Control Equipment: Electrical Fast Transient/Burst Requirements," IEC, American Standards Institute, New York, 1988.

22. JEDEC Standard Number 17, "A Standardized Test Procedure for the Characterization of Latch-up in CMOS Integrated Circuits," Electronic Industries Association, Washington, DC, 1988.

Appendix A

A.1 ERROR MODELS

Error models have an important effect on the computation of aliasing probability. There are two important components of any error model, *temporal* and *spatial*. The first aspect, temporal, models the correlation between the errors caused by different input vectors. For combinational circuits and faults, errors can be assumed to be independent of time. The second aspect, spatial, in multiple-output circuits, is the manner in which the errors manifest themselves at the various outputs of the circuit for any randomly chosen test vector. If the error value for test t_i is e_i for an *n*-input circuit, $e_i = 0$ indicates that t_i did not detect a fault. A nonzero value of e_i indicates that t_i detected a fault and the fault-effect was observed at one or more outputs.

Independent Error Model. This model assumes that the errors at the various outputs of the circuit are independent. The probability, p, represents the detection probability, i.e., the probability that any output of the CUT may be in error. This probability is assumed to be the same for each output. Also, p is assumed to be independent of time, and therefore does not depend on a specific test. The independence of errors with time is realistic in this model, but the independence of space is not. Spatial independence implies that the likelihood of error occurring at one output is more than at multiple outputs. For example, under this model, the probability of any two outputs being in error is $p^2(1 - p)^{n-2}$, which is less than the probability of a single output being in error. In reality, whether the errors at the outputs are independent depends on the circuit structure.

Symmetric Error Model. In any circuit, different outputs share some logic. A fault in the shared logic may cause errors at all the outputs that share the logic. Hence, for an arbitrary circuit, one can assume a certain correlation between errors at all the outputs of the CUT. Modeling the exact nature of the correlation for a given circuit may require exhaustive fault simulation. An approximation to represent the effect of correlation would be to assume that all the $(2^n - 1)$ possible errors are equally likely. Therefore, this assumes that the likelihood of a given error pattern does not depend on the number of bits in error. This model assumes that all $2^n - 1$ erroneous outputs occur with equal probability, $p/(2^n - 1)$. The probability that a given input vector causes an erroneous output is p, the probability that the output is error-free is $(1 - p)$.

General Error Model. Each of the above error models can only be used efficiently for some circuits. However, in general, determining which error model is appropriate for a given circuit may be difficult. To overcome this problem, a general error model is proposed. All time-independent error models are subsumed by this model. This model assumes that the error e_i can take any of its 2^n values $[0, 1, \alpha, \alpha^2, \ldots, \alpha^{2^n-2}]$, with probabilities $p_0, p_1, \ldots, p_{2^n-1}$, respectively.

A.2 COMPUTATION OF ALIASING PROBABILITY

Aliasing probability depends on the error and not on the nature of the fault-free response. In MISR compression, to determine whether a given error sequence will cause aliasing, compressing the error sequence by the MISR is sufficient. If the final state of the MISR is all-zero, the given error sequence will cause aliasing. The aliasing probability in the MISR compression can be computed by different methods. Some are given below.

Equivalent LFSR Compression. This method is based on the linearity of MISR compressors. The input sequence, applied to the inputs $i_0, \ldots, i_{n-1}$ can be replaced by the equivalent sequence applied to the same MISR at input i_0. Thus, the problem of computation of aliasing for MISR compression is replaced by the problem of computation of aliasing for LFSR.

Consider the circuit shown in Figure A.1. The fault-free response for this circuit is also shown in the figure. Figure A.2 shows the equivalent input sequence r_{eq} that can be applied to the LFSR with the same feedback polynomial. Note that this equivalent sequence is obtained by a bit-by-bit XOR operation with the circuit outputs shifted to the left for each test vector. The signature in this case is 010, as before.

If the independent error model is used, the probability of the first bit of the equivalent sequence being in error is p. The probability of the second bit being in error is $2p(1 - p)$, since it is obtained by XOR of two bits in the original response. Hence, the errors in the equivalent input sequence are independent, but their probabilities are not equal. The iterative technique to compute aliasing for LFSR can be used with the equivalent sequence because it does not assume equal bit error rates.

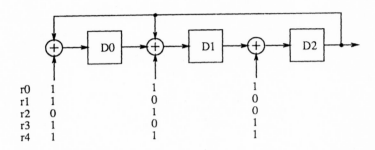

Figure A.1 Example of MISR compression.

Markov Model. The state transition of the MISR depends on the inputs. The initial state of the MISR is considered to be all-zero state. The next state of the MISR will be all-zero if the first output of the CUT (*n*-bit) is error-free. However, it takes $2^n - 1$ nonzero values if the CUT response to the test vector is erroneous. Since the error model defines the probabilities of all the different error patterns, a probability can be assigned to each of the 2^n transitions. The probability of being in any state i, at time t, can be defined as $\pi_i(t)$. The vector $\Pi(t) = [\pi_0(t), \ldots, \pi_{2^n-1}(t)]^T$, is defined as the state probability vector. An error model can be used to define the state transition matrix $\mathbf{A}$. The following recursive relation can be used to compute the state probability matrix at time t:

$$\Pi(t) = \mathbf{A} \cdot \Pi(t - 1)$$

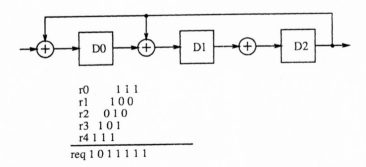

Figure A.2 Equivalent LFSR compression.

This recursive relation defines a Markov process. The aliasing probability for a test length t is given by

$$P_{aliasing} = \pi_0(t) - (1 - p)^n t$$

This aliasing probability is the probability that the MISR is in the all-zero state (in the error domain) after compressing t test responses. The $(1 - p)^t$ term precludes the case that all circuit output responses are error-free.

Coding Theory. Coding theory formulation is widely used to study aliasing in MISR compression. The advantage is that the exact closed-form expressions for aliasing probability can be obtained under the symmetric error model for any test length, as well as under the independent error model for some test lengths. Closed-form expressions for MISR aliasing under the independent error model remain an open problem.

The response of the circuit is commonly compressed by an n-stage MISR with a primitive polynomial $\phi(x)$. This polynomial has an equivalent representation as a one-degree feedback polynomial, $\phi(x) = x + \alpha$ over $GF(2^n)$, as shown in Figure A.3, where α is the primitive element in the $GF(2^n)$ generated by using $\phi(x)$ as a generator ($\alpha^i \neq \alpha^j$, for $i \neq j$; $i, j = 0, 1, \ldots, 2^n - 2$). Consider an n-output circuit. Let N be the number of test vectors and $R(x)$ be the fault-free response. $R(x)$ is given as

$$R(x) = r_{N-1}x^{N-1} + r_{N-2}x^{N-2} + \cdots + r_1 x + r_0$$

If $R'(x)$ is a faulty response, it is given as

$$R'(x) = r'_{N-1}x^{N-1} + r'_{N-2}x^{N-2} + \cdots + r'_1 x + r'_0$$

Let $E(x)$ be the error polynomial, where $E(x)$ is given as

$$E(x) = e_{N-1}x^{N-1} + e_{N-2}x^{N-2} + \cdots + e_1 x + e_0$$

Note that $E(x) = R(x) + R'(x)$, and all additions are modulo-2. For LFSR compression with feedback polynomial $\phi(x)$, the fault-free signature $S(x)$ is given by

$$R(x) = Q(x) \cdot \phi(x) + S(x)$$

Similarly, the faulty signature $S'(x)$ is given by

$$R'(x) = Q'(x) \cdot \phi(x) + S'(x)$$

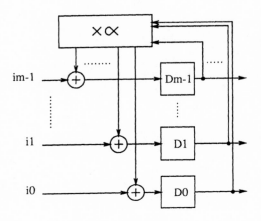

Figure A.3 Equivalent representation of an *n*-stage MISR

Note that degree of $S(x)$ and $S'(x)$ is less than the degree of $\phi(x)$. Aliasing occurs when the faulty response $R'(x) \neq R(x)$, but $S'(x) = S(x)$. Due to the linearity of these compressors, the fault-free and faulty signatures are identical, if and only if the corresponding error polynomial $E(x)$ is divisible by $\phi(x)$. The set of all polynomials divisible by $\phi(x)$ constitutes a code with generator polynomial $\phi(x)$. This code is called the *aliasing code* (AC) of the compressor. The AC for MISR is a distance-2 code over $GF(2^n)$. If the degree of $\phi(x)$ is 1, the aliasing code is MDS. If N tests are applied, it is a $(N, N - 1)$ MDS code. These codes correspond to the Reed-Solomon codes when $N = 2^n - 1$.

Consider a two-output circuit on which three test vectors are applied ($N = 3$). Let the response be compressed by an MISR with feedback polynomial $\phi(x) = x^2 + x + 1$ (Figure A.4(a)). Note that this MISR can be represented as a single stage MISR over $GF(2^2)$ with $\phi(x) = x + \alpha$, as shown in Figure A.4b, where α is the primitive element in $GF(2^2)$ generated by $\phi(x) = x^2 + x + 1$, i.e., $\alpha^2 + \alpha + 1 = 0$. In this case, the field has four elements $0 = (0,0)$, $1 = (0,1)$, $\alpha = (1,0)$, and $\beta = (1,1)$, where $\beta = \alpha^2 + \alpha + 1$. Only those $R'(x)$ will cause aliasing for which the error E belongs to the aliasing code, given as follows:

$$
AC = \begin{bmatrix}
000 & 1\alpha0 & \beta0\alpha & \beta\alpha1 \\
01\alpha & \alpha\beta0 & 10\beta & 111 \\
0\alpha\beta & \beta10 & 1\beta\alpha & \alpha\alpha\alpha \\
0\beta1 & \alpha01 & \alpha1\beta & \beta\beta\beta
\end{bmatrix}
$$

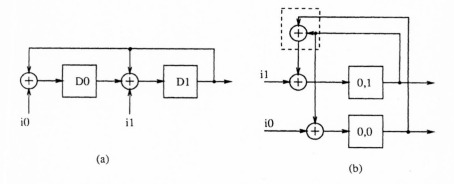

Figure A.4 MISR with (a) primitive feedback polynomial and (b) equivalent single stage representation.

Thus, in this case, out of $4^3 - 1 = 63$ possible nonzero error vectors, 15 nonzero errors will result in aliasing. Key to the developing expressions for aliasing probability is the formulation of expressions for appropriate weight enumerators for the aliasing code. There are different weight enumerators, the type of weight enumerator to be used depending on the particular error model. Hamming weight enumerators for the distance-2 Reed-Solomon code can be used to compute the aliasing probability for the symmetric error model. For any test length, this is given by

$$P_{aliasing} = \frac{1}{2^n} \cdot \left[1 - 2^n(1 - p)^N + (2^n - 1)\left(1 - \frac{2^n p}{2^n - 1}\right)^N \right]$$

The aliasing probability for the independent error model can be computed by using the weight enumerator of the binary image of the aliasing code. If $N = r(2^n - 1)$ tests are applied to an n-output circuit, the output of which is compressed by an MISR with primitive feedback polynomial, the aliasing probability under independent error model is given by

$$P_{aliasing} = \frac{1}{2^n} \cdot \left[1 - (2^n - 1)(1 - 2p)^{r_n(2^n/2)} \right] - (1 - p)^{r_n(2^n - 1)}$$

Note that for large test lengths, all error models converge to the same aliasing probability. However, for small N, the independent error model predicts maximum aliasing, whereas the symmetric error model predicts minimum aliasing. Also, note that if the circuit has small number of outputs, all error models give similar results.

Appendix B
Annotated Bibliography

BOOKS

1. M.A. Breuer and A.D. Friedman, *Diagnosis and Reliable Design of Digital Systems*, Computer Science Press, W.H. Freeman, N.Y., 1976.
2. J.P. Roth, *Computer Logic Testing and Verification*, Computer Science Press, W.H. Freeman, N.Y., 1980.
3. D.P. Siewiorek and R.S. Swarz, *The Theory and Practice of Reliable System Design*, Digital Press, 1982.
4. R.G. Bennetts, *Design of Testable Logic*, Addison Wesley, Reading, MA, 1984.
5. H. Fujiwara, *Logic Testing and Design for Testability*, MIT Press, Cambridge, MA, 1985.
6. P.K. Lala, *Fault Tolerant and Fault Testable Hardware Design*, Prentice Hall, Englewood Cliffs, NJ, 1985.
7. A. Miczo, *Digital Logic Testing and Simulation*, Harper and Row, New York, 1986.
8. D.K. Pradhan, *Fault Tolerant Computing: Theory and Practice*, Prentice Hall, Englewood Cliffs, NJ, 1986.
9. F.F. Tsui, *LSI/VLSI Testability Design*, McGraw-Hill, New York, 1987.
10. J.M. Cortner, *Digital Test Engineering*, John Wiley and Sons, New York, 1987.
11. P.H. Bardell, W.H. McAnney, and J. Savir, *Built-in Test for VLSI*, John Wiley and Sons, New York, 1987.
12. R.J. Feugate and S.M. McIntyre, *Introduction to VLSI Testing*, Prentice Hall, Englewood Cliffs, NJ, 1988.
13. R.E. Massara, *Design and Test Techniques for VLSI and WSI Circuits*, Peter Peregrinus, London, 1989.

14. I. Koren, *Defect and Fault Tolerance in VLSI Systems*, Plenum Press, New York, 1989.
15. M. Abramovici, M.A. Breuer, and A.D. Friedman, *Digital System Testing and Testable Design*, Computer Science Press, W.H. Freeman, N.Y., 1990.
16. V.N. Yarmolik, *Fault Diagnosis of Digital Circuits*, John Wiley and Sons, New York, 1990.
17. E.B. Eichelberger, E. Lindbloom, J. Waicukauski, and T.W. Williams, *Structured Logic Testing*, Prentice Hall, Englewood Cliffs, NJ, 1991.
18. W. Needham, *Designer's Guide to Testable ASIC Devices*, Van Nostrand Reinhold, New York, 1991.

JOURNALS

1. *IEEE Transactions on Computers.*
2. *IEEE Transactions on CAD of Integrated Circuits and Systems.*
3. *IEEE Transactions on Circuits and Systems.*
4. *IEEE Transactions on Reliability.*
5. *IEEE Journal Solid State Circuits.*
6. *IEEE Transactions Industrial Electronics.*
7. *IEEE Design and Test of Computers.*
8. *IEEE Computer.*
9. *IEEE Circuits and Devices.*
10. *IEE Proceedings, Part E.*
11. *IEE Proceedings, Part G.*
12. *Electronics Letters*, IEE.
13. *International Journal Electronic Testing*, Kluwer.
14. *VLSI System Design*, CMP Publications.
15. *Test and Measurement World*, Cahners Publications.
16. *Bell System Technical J.*

CONFERENCE PROCEEDINGS AND RECORDS

1. International Test Conference.
2. International Conference on Computer Design.
3. Design Automation Conference.
4. International Conference on Computer Aided Design.
5. International Symp. on Fault Tolerant Computing.
6. International Symp. on Circuits and Systems.
7. IEEE Custom Integrated Circuits Conference.
8. IEEE VLSI Test Symposium.
9. IEEE Reliability Physics Symposium.

10. European Design Automation Conference.
11. European Test Conference.
   In addition to the above lists, papers related to testing have appeared in other VLSI design and computer architecture conferences and journals. There also are many workshops held annually, which are not listed here.

ADDITIONAL SUGGESTED READING

The papers listed below are additional references on various topics, which have not been referenced at the end of a chapter. Many more papers have been published. The list given below contains only few a selected papers.

1. H. Fujiwara and S. Toida, "The Complexity of Fault Detection Problem for Combinational Logic Circuits," *IEEE Trans. Comp.*, **31**(6), pp. 555–560, June 1982.
2. L.H. Goldstein, "Controllability/Observability Analysis of Digital Circuits," *IEEE Trans. Circuits and Sys.*, **26**(9), pp. 685–693, Sep. 1979.
3. J. Grason, "TMEAS, A Testability Measurement Program," *Proc. Design Auto. Conf.*, pp. 156–161, 1979.
4. S.C. Seth, V.D. Agarwal, and H. Farhat, "A Statistical Theory of Digital Circuit Testability," *IEEE Trans. Comp.*, **39**(4), pp. 582–586, April 1990.
5. J.A. Abraham and W.K. Fuchs, "Fault and Error Models for VLSI," *Proc. IEEE*, **74**(5), pp. 639–654, May 1986.
6. R.R. Fritzemeier, H.T. Nagle, and C.F. Hawkins, "Fundamentals of Testability—A Tutorial," *IEEE Trans. Industrial Electronics*, **36**(2), pp. 117–128, May 1989.
7. D.C.Y. Mei, "Bridging and Stuck-at Faults," *IEEE Trans. Comp.*, **33**(7), pp. 720–727, July 1974.
8. A. Iosupovicz, "Optimal Detection of Bridging Faults and Stuck-at Faults in Two Level Logic," *IEEE Trans. Computers*, **27**(5), pp. 452–455, May 1978.
9. M. Karpovsky and S.Y.H. Su, "Detection and Location of Input and Feedback Bridging Faults among Input and Output Lines," *IEEE Trans. Comp.*, **29**(6), pp. 523–527, June 1980.
10. K.L. Kodandapani and D.K. Pradhan, "Undetectability of Bridging Faults and Validity of Stuck-at Fault Tests," *IEEE Trans. Comp.*, **29**(1), pp. 55–59, Jan. 1980.
11. S. Xu and S.Y.H. Su, "Testing Feedback Bridging Faults among Internal Input and Output Lines by Two Patterns," *Proc. Int. Conf. Circuits and Computers*, pp. 214–217, 1982.
12. Y.M. El-Ziq and S.Y.H. Su, "Fault Diagnosis of MOS Combinational Networks," *IEEE Trans. Comp.*, **31**(2), pp. 129–139, Feb. 1982.
13. P.Benerjee and J.A. Abraham, "Generating Tests for Physical Failures in MOS Logic Circuits," *Proc. Int. Test Conf.*, pp. 554–559, Oct. 1983.

14. M. Karpovsky, "Universal Tests for Detection of Input/Output Stuck-at and Bridging Faults," *IEEE Trans. Computers*, **32**(12), pp. 1194–1198, Dec. 1983.

15. S. Xu and S.Y.H. Su, "Detecting I/O and Internal Feedback Bridging Faults," *IEEE Trans. Computers*, **34**(6), pp. 553–557, June 1985.

16. B.B. Bhattacharya and B. Gupta, "On the Impossible Class of Faulty Functions in Logic Networks under Short Circuit Faults," *IEEE Trans. Comp.*, **35**, pp. 85–90, Jan. 1986.

17. R. Rajsuman, Y.K. Malaiya, and A.P. Jayasumana, "Limitations of Switch Level Analysis for Bridging Faults," *IEEE Trans. CAD*, **8**(7), pp. 807–811, July 1989.

18. S.D. Millman, E.J. McCluskey, and J.M. Acken, "Diagnosing CMOS Bridging Faults with Stuck-at Fault Dictionaries," *Proc. Int. Test Conf.*, pp. 860–870, 1990.

19. M.E. Zaghloul and D. Gobovic, "Fault Modeling of Physical Failures in CMOS VLSI Circuits," *IEEE Trans. Cir. and Sys.*, **37**(12), pp. 1528–1543, Dec. 1990.

20. Y.M. El-Ziq and R.J. Cloutier, "Functional Level Test Generation for Stuck-Open Faults in CMOS VLSI," *Proc. Int. Test Conf.*, pp. 536–546, 1981.

21. R. Chandramouli, "On Testing Stuck-Open Faults," *Proc. Int. Symp. Fault Tolerant Computing*, pp. 258–265, 1983.

22. S.M. Reddy, M.K. Reddy, and V.D. Agarwal, "Robust Tests for Stuck-Open Faults in CMOS Combinational Logic Circuits," *Proc. Int. Symp. Fault Tol. Comp.*, pp. 44–49, 1984.

23. P.S. Moritz and L.M. Thorsen, "CMOS Circuit Testability," *IEEE J. Solid State Cir.*, **21**(4), pp. 306–309, April 1986.

24. N. Ling and M.A. Bayoumi, "An Efficient Technique to Improve NORA CMOS Testing," *IEEE Trans. Cir. and Sys.*, **34**(12), pp. 1609–1611, Dec. 1987.

25. S. Koeppe, "Optimal Layout to Avoid CMOS Stuck-Open Faults," *Proc. 24th Design Auto. Conf.*, pp. 687–691, 1987.

26. S.D. Sherlekar and P.S. Subramaian, "Conditionally Robust Two Pattern Tests and CMOS Design for Testability," *IEEE Trans. CAD*, **7**(3), pp. 325–332, March 1988.

27. N.K. Jha, "Multiple Stuck-Open Fault Detection in CMOS Logic Circuits," *IEEE Trans. Comp.*, **37**, pp. 426–432, April 1988.

28. J.M. Soden, R.K. Treece, M.R. Taylor, and C.F. Hawkins, "CMOS IC Stuck-Open Fault Electrical Effects and Design Considerations," *Proc. Int. Test Conf.*, pp. 423–430, 1989.

29. R. Rajsuman, A.P. Jayasumana, and Y.K. Malaiya, "CMOS Stuck-Open Fault Detection Using Single Test Patterns," *Proc. IEEE/ACM Design Auto. Conf.*, pp. 714–717, 1989.

30. C. Landrault and S. Pravossoudovitch, "Hazards Effect on Stuck-Open Fault Testability," *Proc. European Test Conf.*, pp. 201–207, 1989.
31. N. Kanopoulos and N. Vasanthavada, "Testing of Differential Cascode Voltage Switch (DCVS) Circuits," *IEEE J. Solid State Circuits*, **25**(3), pp. 806–813, June 1990.
32. S.K. Jain and V.D. Agarwal, "Modeling and Test Generation Algorithms for MOS Circuits," *IEEE Trans. Comp.*, **34**(5), pp. 426–433, May 1985.
33. R.E. Brayant, "A Switch Level Model and Simulator for MOS Digital Systems," *IEEE Trans. Comp.*, **33**(2), pp. 160–177, Feb. 1984.
34. C.F. Chen, C.Y. Lo, H.N. Nham, and P. Subramaniam, "The Second Generation Motis Mixed Mode Simulator," *Proc. Design Auto. Conf.*, pp. 10–17, 1984.
35. M. Kawai and J.P. Hayes, "An Experimental MOS Fault Simulation Program, CSASIM," *Proc. Design Auto. Conf.*, pp. 2–9, 1984.
36. T. Ibaraki, T. Kameda, and S. Toida, "On Minimal Test Sets for Locating Single Link Failures in Networks," *IEEE Trans. Comp.*, **30**(3), pp. 182–189, March 1981.
37. V.K. Agarwal and A.S.F. Fung, "Multiple Fault Testing of Large Circuits by Single Fault Test Sets," *IEEE Trans. Comp.*, **30**(11), pp. 855–865, Nov. 1981.
38. M.A. Heap and W.A. Rogers, "Generating Single Stuck-at Fault Coverage from a Collapsed Fault Set," *IEEE Computer*, pp. 51–57, April 1989.
39. J. Khakbaz and E.J. McCluskey, "Concurrent Error Detection and Testing for Large PLAs," *IEEE J. Solid State Cir.*, **17**, pp. 386–394, 1982.
40. K.S. Ramanatha and N.N. Biswas, "An On-line Algorithm for the Location of Cross Point Faults in Programmable Logic Arrays," *IEEE Trans. Comp.*, **32**(5), pp. 438–444, May 1983.
41. K.S. Ramanatha and N.N. Biswas, "A Design for Testability of Undetectable Crosspoint Faults in PLAs," *IEEE Trans. Comp.*, **32**(6), pp. 554–557, June 1983.
42. Y. Tamir and C.H. Sequin, "Design and Application of Self Testing Comparators Implemented with MOS PLAs," *IEEE Trans. Comp.*, **33**(6), pp. 493–506, June 1984.
43. J. Khakbaz, "A Testable PLA Design with Low Overhead and High Fault Coverage," *IEEE Trans. Comp.*, **33**(8), pp. 743–745, Aug. 1984.
44. M.A. Breuer and X. Zhu, "A Knowledge Based System for Selecting a Test Methodology for a PLA," *Proc. Design Auto. Conf.*, pp. 259–266, 1985.
45. R.S. Wei and A.S. Vincentelli, "PLATYPUS: A PLA Test Pattern Generation Tool," *Proc. Design Auto. Conf.*, pp. 197–204, 1985.
46. M.M. Lighthart, E.H.L. Aarts, and F.P.M. Beenker, "PLA Design for Testability Using Statistical Cooling," *Proc. Design Auto. Conf.*, pp. 339–345, 1986.

47. H.K. Reghbati, "Fault Detection in PLAs," *IEEE Design and Test*, **3**(4), pp. 43–50, Dec. 1986.

48. C.L. Wey, M.K. Vai, and F. Lombardi, "On the Design of Redundant Programmable Logic Array (RPLA)," *IEEE J. Solid State Cir.*, **22**, pp. 114–117, Feb. 1987.

49. T.Y. Chang and C.L. Wey, "Design of Fault Diagnosable and Repairable PLAs," *IEEE J. Solid State Circuits*, **24**(5), pp. 1451–1454, Oct. 1989.

50. S.J. Upadhyaya and J.A. Thodiyil, "BIST PLAs, Pass or Fail—A Case Study," *Proc. Design Auto. Conf.*, pp. 724–727, 1990.

51. H. McAdams, J.H. Neal, B. Holland, S. Inoue, W.K. Loh, and K. Poteet, "A 1-Mbit CMOS Dynamic RAM with Design for Test Functions," *IEEE J. Solid State Cir.*, **21**(5), pp. 635–641, Oct. 1986.

52. T. Ohsawa et al., "A 60-ns 4-Mbit CMOS DRAM with Built-in Self-Test Function," *IEEE J. Solid State Cir.*, **22**(5), pp. 663–667, Oct. 1987.

53. P.H. Voss *et al.*, "A 14-ns 256K × 1 CMOS SRAM with Multiple Test Modes," *IEEE J. Solid State Cir.*, **24**(4), pp. 874–880, Aug. 1989.

54. R. David, A. Fuentes, and B. Courtois, "Random Pattern Testing versus Deterministic Testing of RAMs," *IEEE Trans. Comp.*, **38**(5), pp. 637–650, May 1989.

55. M.F. Chang, W.K. Fuchs, and J.H. Patel, "Diagnosis and Repair of Memory with Coupling Faults," *IEEE Trans. Comp.*, **38**(4), pp. 493–500, April 1989.

56. P. Mazumder and H.J. Patel, "Parallel Testing for Pattern Sensitive Faults in Semiconductor Random Access Memories," *IEEE Trans. Comp.*, **38**(3), pp. 394–407, March 1989.

57. T. Fuja, C. Heegard, and R. Goodman, "Linear Sum Codes for Random Access Memories," *IEEE Trans. Comp.*, **37**(9), pp. 1030–1042, Sep. 1988.

58. R. David and A. Fuentes, "Fault Diagnosis of RAMs from Random Testing Experiments," *IEEE Trans. Comp.*, **39**(2), pp. 220–229, Feb. 1990.

59. C.L. Wey and F. Lombardi, "On the Repair of Redundant RAMs," *IEEE Trans. CAD*, **6**(2), pp. 222–231, March 1987.

60. M.A. Breuer, "A Random and an Algorithmic Technique for Fault Detection Test Generation for Sequential Circuits," *IEEE Trans. Comp.*, **20**(11), pp. 1364–1370, Nov. 1971.

61. F.J. Hill and B. Huey, "Scirtss: A Search System for Sequential Circuit Test Sequences," *IEEE Trans. Comp.*, **26**(5), pp. 490–502, May 1977.

62. S.I. Murakami, K. Kinoshita, and H. Ozaki, "Sequential Machines Capable of Fault Diagnosis," *IEEE Trans. Comp.*, **19**(11), pp. 1079–1085, Nov. 1970.

63. M.C. Browne, E.M. Clarke, D.L. Dill, and B. Mishra, "Automatic Verification of Sequential Circuits Using Temporal Logic," *IEEE Trans. Comp.*, **35**(12), pp. 1035–1043, Dec. 1986.

64. S. Devadas, H.T. Ma, and A.R. Newton, "On the Verification of Sequential Machines at Differing Levels of Abstraction," *IEEE Trans. CAD*, **7**(6), pp. 713–722, June 1988.

65. H.T. Ma, S. Devadas, A.R. Newton, and A.S. Vincentelli, "Test Generation for Sequential Circuits," *IEEE Trans. CAD*, **7**(10), pp. 1081–1091, Oct. 1988.

66. R.W. Cook, W.H. Sisson, T.F. Storey, and W.N. Toy, "Design of a Self-Checking Microprogram Control," *IEEE Trans. Comp.*, **22**(3), pp. 255–262, March 1973.

67. L. Shen and S.Y.H. Su, "A Functional Testing Method for Microprocessors," *IEEE Trans. Comp.*, **37**(10), pp. 1288–1293, Oct. 1988.

68. H.T. Nagle, R.R. Fritzemeier, J.E. Van Well, and M.G. Mcnamer, "Microprocessor Testability," *IEEE Trans. Indust. Elect.*, **36**(2), pp. 151–163, May 1989.

69. M.G. Gallup *et al.*, "Testability Features of the 68040," *Proc. Int. Test Conf.*, pp. 749–757, 1990.

70. F. Muradali, V.K. Agarwal, and B.N. Dostie, "A New Procedure for Weighted Random Built-in Self-Test," *Proc. Int. Test Conf.*, pp. 660–669, 1990.

71. M. Damiani *et al.*, "Aliasing in Signature Analysis Testing with Multiple Input Shift Registers," *IEEE Trans. CAD*, **9**(12), pp. 1344–1353, Dec. 1990.

72. R. David, "Testing by Feedback Shift Register," *IEEE Trans. Comp.*, **29**(7), pp. 668–673, July 1980.

73. J.P. Hayes, "Transition Count Testing of Combinational Logic Circuits," *IEEE Trans. Comp.*, **25**(6), pp. 613–620, June 1976.

74. K. Iwasaki, "Analysis and Proposal of Signature Circuits for LSI Testing," *IEEE Trans. CAD*, **7**(1), pp. 84–90, Jan. 1988.

75. K. Iwasaki and F. Arakawa, "An Analysis of the Aliasing Probability of Multiple Input Signature Registers in the Case of a $2m$-ary Symmetric Channel," *IEEE Trans. CAD*, **9**(4), pp. 427–438, April 1990.

76. G. Markowski, "Syndrome Testability Can be Achieved by Circuit Modification," *IEEE Trans. Comp.*, **30**(8), pp. 604–608, Aug. 1981.

77. D.K. Pradhan, S.K. Gupta, and M.G. Karpovsky, "Aliasing Probability for Multiple Input Signature Analyzer and a New Compression Technique," *IEEE Trans. Comp.*, **39**(4), pp. 586–591, April 1990.

78. J.P. Robinson and N.R. Sexena, "A Unified View of Test Compression Methods," *IEEE Trans. Comp.*, **36**(1), pp. 94–99, Jan. 1987.

79. J.E. Smith, "Measures of the Effectiveness of Fault Signature Analysis," *IEEE Trans. Comp.*, **29**(6), pp. 510–514, June 1980.

80. L. Ward and E.J. McCluskey, "Condensed Linear Feedback Shift Register (LFSR) Testing—A Pseudoexhaustive Test Technique," *IEEE Trans. Comp.*, **35**(4), pp. 367–370, April 1986.

81. K.D. Wagner, C.K. Chin, and E.J. McCluskey, "Pseudorandom Testing," *IEEE Trans. Comp.*, **36**(3), pp. 332–343, March 1987.

82. M.E. Turner, D.G. Leet, R.J. Prilik, and D.J. Mclean, "Testing CMOS VLSI: Tools, Concepts, and Experimental Results," *Proc. Int. Test Conf.*, pp. 322–328, 1985.

83. J.F. Frenzel and P.N. Marinos, "Power Supply Current Signature (PSCS) Analysis: A New Approach to System Testing," *Proc. Int. Test Conf.*, pp. 125–129, 1987.
84. D. Burns, "Locating High Resistance Shorts in CMOS Circuits by Analyzing Supply Current Measurement Vectors," *Int. Symp. Test and Failure Analysis*, pp. 231–237, 1989.
85. T.M. Storey and W. Maly, "CMOS Bridging Fault Detection," *Proc. Int. Test Conf.*, pp. 842–850, 1990.
86. P. Nigh and W. Maly, "Test Generation for Current Testing," *IEEE Design and Test of Computers*, pp. 26–38, Feb. 1990.
87. Y.K. Malaiya, A.P. Jayasumana, Q. Tong, and S.M. Menon, "Enhancement of Resolution in Supply Current Based Testing for Large ICs," *Proc. IEEE VLSI Test Symp.*, pp. 291–296, 1991.
88. K.J. Lee and M.A. Breuer, "Constraints for Using IDDQ Testing to Detect CMOS Bridging Faults," *Proc. IEEE VLSI Test Symp.*, pp. 303–308, 1991.
89. C.H. Chen and J.A. Abraham, "High Quality Tests for Switch Level Circuits Using Current and Logic Test Generation Algorithms," *Proc. Int. Test Conf.*, 1991.
90. E. Vandris and G. Sobelman, "A Mixed Functional/IDDQ Testing Methodology for CMOS Transistor Faults," *Proc. Int. Test Conf.*, 1991.
91. S.K. Malik and E.F. Chace, "MOS Gate Oxide Quality Control and Reliability Assessment by Voltage Ramping," *Proc. Int. Test Conf.*, pp. 384–389, 1984.
92. F.H. Hielscher and J.C. Pagano, "Backdrive Stress Testing of CMOS Gate Array Circuits," *Proc. Int. Test Conf.*, pp. 523–533, 1985.
93. J.H. Hendriks, "Overdriving n-MOS and CMOS VLSI Circuits," *Proc. Int. Test Conf.*, pp. 534–539, 1985.
94. D.A. Angst and J. Domitrowich, "Trends in Parametric Test Systems," *Semiconductor International*, pp. 173–177, Sep. 1987.
95. M.G. Mcnamer, S.C. Roy, and H.T. Nagle, "Statistical Fault Sampling," *IEEE Trans. Indust. Elect.*, **36**(2), pp. 141–150, May 1989.
96. W.D. Ballew and L.M. Streb, "Incoming Test Strategy Based Upon In-Process Failure and Repair Costs," *IEEE Trans. Indust. Elect.*, **36**(2), pp. 203–210, May 1989.
97. E.J. McCluskey and F. Buelow, "IC Quality and Test Transparency," *IEEE Trans. Indust. Elect.*, **36**(2), pp. 197–202, May 1989.
98. R. Menozzi *et al.*, "Latch-up Testing in CMOS ICs," *IEEE J. Solid State Cir.*, **25**(4), pp. 1010–1014, Aug. 1990.
99. J.M. Sweet, M.R. Tuck, D.W. Peterson, and D.W. Palmer, "Short and Long Loop Manufacturing Feedback Using a Multisensor Assembly Test Chip," *IEEE Trans. CHMT*, **14**(3), pp. 529–535, Sep. 1991.
100. N.K. Jha and J.A. Abraham, "Design of Testable CMOS Logic Circuits under Arbitrary Delays," *IEEE Trans. on CAD*, **4**(7), pp. 264–269, July 1985.

Index

The Artech House Telecommunications Library

Vinton G. Cerf, Series Editor